EUROPEAN ORIGINS OF LIBRARY AND INFORMATION SCIENCE

STUDIES IN INFORMATION

Series Editor: **Jens-Erik Mai**

Recent Volumes (Library and Information Science)

Dania Bilal and Jamshid Beheshti
New Directions in Children's and Adolescents' Information Behavior Research

Christine Bruce, Kate Davis, Hilary Hughes, Helen Partridge, and Ian Stoodley
Information Experience: Approaches to Theory and Practice

Mark Hepworth and Geoff Walton
Developing People's Information Capabilities: Fostering Information Literacy in Educational, Workplace and Community Contexts

Jung-Ran Park and Lynne C. Howarth
New Directions in Information Organization

Amanda Spink and Jannica Heinström
Trends and Research: Europe

Gunilla Widén and Kim Holmberg
Social Information Research

Dirk Lewandowski
Web Search Engine Research

Donald Case
Looking for Information, Third Edition

Amanda Spink and Diljit Singh
Trends and Research: Asia-Oceania

Amanda Spink and Jannica Heinstrom
New Directions in Information Behaviour

Eileen G. Abels and Deborah P. Klein
Business Information: Needs and Strategies

Leo Egghe
Power Laws in the Information Production Process: Lotkaian Informetrics

Mike Thelwall
Link Analysis: An Information Science Approach

Matthew Locke Saxton and John V. Richardson
Understanding Reference Transactions: Transforming an Art into a Science

Robert M. Hayes
Models for Library Management, Decision-Making, and Planning

Charles T. Meadow, Bert R. Boyce, and Donald H. Kraft
Text Information Retrieval Systems, Second Edition

Jack Andersen
Genre Theory in Information Studies

Jack Andersen and Laura Skouvig
The Organization of Knowledge

STUDIES IN INFORMATION

EUROPEAN ORIGINS OF LIBRARY AND INFORMATION SCIENCE

BY

FIDELIA IBEKWE
School of Journalism & Communication, Aix-Marseille University, Marseille, France

Series Editor: Jens-Erik Mai

emerald PUBLISHING

United Kingdom – North America – Japan
India – Malaysia – China

Emerald Publishing Limited
Howard House, Wagon Lane, Bingley BD16 1WA, UK

First edition 2019

British Library Cataloguing in Publication Data
A catalogue record for this book is available from the British Library

ISBN: 978-1-78756-718-4 (Print)
ISBN: 978-1-78756-717-7 (Online)
ISBN: 978-1-78756-719-1 (EPub)

ISSN: 2055-5377 (Series)

Printed and bound by CPI Group (UK) Ltd, Croydon, CR0 4YY

ISOQAR certified Management System, awarded to Emerald for adherence to Environmental standard ISO 14001:2004.

ISOQAR REGISTERED
Certificate Number 1985
ISO 14001

INVESTOR IN PEOPLE

Acknowledgements

The field work on this multi-country research would not have been possible without the financial support from the 2016 History Fund of the Association for Information Science and Technology which enabled me to travel to Croatia, Spain and Denmark in order to interview Library and Information Science (LIS) scholars.

I owe special thanks to Tatjana Aparac Jelušić, Virginia Ortiz-Repiso Jimenez and Niels Lund Windfield, who smoothed my path to interviewees in their respective countries. Their hospitality and readiness in contacting their colleagues and in organising the interviews ensured that everything went smoothly during my stay. I also thank them for providing documentary materials on their pioneers and for reading earlier drafts of the history on their country. Likewise, my heartfelt gratitude to all the interviewees, for without them, it would have been impossible to obtain some undocumented information on the historic context of the development of the field in their country. They are by order of appearance of the chapters in the book: Tatjana Aparac Jelusic and Neva Tudor-Šilović (for ex-Yugoslavia), Niels Windfield Lund (for Denmark, Sweden and Norway), Ernest Abadal, Tomás Baiget, Agustí Canals, Virginia Ortiz-Repiso Jimenez, José Antonio Moreiro, Maria-Antonia García Moreno, Fernando Ramos, Mercedes Caridad Sébastian (for Spain) and Fernanda Ribeiro (for Portugal).

Let me also acknowledge the invaluable help of Thomas Dousa, librarian *extraordinaire* at the University of Chicago Library, in tracking down obscure documentary resources and providing access to publications that were outside my reach.

Several other colleagues read drafts of this book especially as it pertained to the origins of LIS in their own country. The chapter on Nordic LIS owes much to Birger Hjørland for his painstaking attention to details and corrections to earlier drafts. I am also indebted to Peter Ingwersen, Louise Limberg and Ragnar Andreas Audunson who also corrected earlier drafts of this chapter and provided supplementary documentary evidence.

Finally, this book owes much to Michael Buckland, Emeritus Professor at the iSchool of Berkeley, California (USA). Our serendipitous meeting in a hotel lobby during the ISKO conference in Montreal in 2008 has proved to be very fruitful. His passion for the European pioneers of documentation and Information science, in particular for the French '*Madame documentation*'. Suzanne Briet proved to be highly contagious and has led me to venture into what was until then unknown territory for me: the history and foundations of LIS. The numerous citations to him in this book are a testimony of his unfailing generosity in sharing his current research work, for reading earlier drafts of this manuscript and for gently pointing out errors, omissions and useful additions to the book.

The final result certainly falls short of the time, efforts and encouragements I received from all these scholars who guided my faltering steps into a cross-country history of LIS. I was not able to take into account all of the suggestions I received. Therefore, any remaining errors, omissions and imprecisions are entirely my responsibility as author. My hope is that the finished product, imperfect as it is, will sow the seeds of a promising plant that other more resourceful scholars across the world will nurture.

Contents

List of Figures

List of Tables

List of Abbreviations

ASIST	Association for Information Science and Technology
JASIST	Journal of the American Society for Information Science & Technology
ASLIB	Association for Special Libraries and Information Bureaux
LIS	Library and Information Science
IS	Information science
IIB	International Institute of Bibliography
IID	Institut International de Documentation
IFLA	International Federation of Library Associations
STI	Scientific and Technical Information
ABF	Association des Bibliothécaires Français
ADBS	Association des Documentalistes et des Bibliothécaires Spécialisés
BNF	Bibliothèque Nationale de France
CNAM	Conservatoire National des Arts et des Métiers
CNRS	Centre National de Recherche Scientifique
ENSSIB	Ecole Nationale Supérieure des Sciences de l'Information et des Bibliothèques
FID	Fédération Internationale de Documentation
INTD	Institut National des Techniques Documentaires
UFOD	Union Française des Organismes de Documentation
CCL	Central Chemical Library
CSLDIS	Center for the Study of Librarianship, Documentation and Information Sciences
ISIP	Internacionalna Stalna Izložba Publikacija (International Permanent Exhibition of Publications)
RC	Referral Centre
RSL	Royal School of Librarianship
RSLIS	Royal School of Library and Information Science
SSLS	Swedish State Library School
SSLIS	Swedish School of Library and Information Science
SRC	Swedish Research Council
NORDINFO	Nordic Council for Scientific Information
NORSLIS	Nordic Research School in Information Studies
NSOL	Norwegian School of Librarians

ESD	Escuela Superior de Diplomática
EB	Escola de Bibliotecàries
SEDIC	Sociedad Española de Documentación e Información Científica
ANABAD	Asociación Nacional de Archiveros, Bibliotecarios, Museólogos y Documentalistas
CSIC	Consejo Superior de Investigaciones Científicas

About the Author

Fidelia Ibekwe obtained a BA in Foreign Languages and Literatures from the University of Port-Harcourt in Nigeria (her native country). She then moved to France to pursue her post-graduate education, obtaining a MA in French Literature (*Lettres Modernes*) from Stendhal University in Grenoble. She then moved to Information and Communication Sciences where she obtained her PhD and her *Habilitation à Diriger des recherches*. She has previously been Associate Professor of Information and Communication Sciences at the University of Nancy 2 and at Jean Moulin University in Lyon. She is currently a Full Professor of Information and Communication Sciences at the School of Journalism and Communication, Aix-Marseille University in France. Her research interests span theoretical, historical and empirical topics. She has worked and published research articles on text mining, terminology extraction and modelling, automatic indexing and information retrieval using natural language processing, statistical and probabilistic models. She is currently researching various issues relating to the history and theoretical foundations of information science, the uses of information and communication technologies and the impact of open and big data on science and on society.

Scope and Aims of the Book

'Study the past if you would define the future.'
– Confucius

The term Library and Information Science (LIS) will be used loosely in this book to refer to the academic field of teaching and enquiry which includes documentation, information science, librarianship and archival studies. Information science (IS) will only be used when the library component is explicitly excluded and the focus is on information. Also, the word 'science' will be used here in the more broader continental European sense of the German '*Wissenschaft*' rather than in the more restrictive British English sense where 'science' denotes the natural and physical sciences.

From a wider historical perspective, the origins of what became known in the twentieth century as Library and Information Science (LIS) is a *continuum* traversing several millennia. Shera and Cleveland (1985, p. 249) recalled that as far back as two centuries before the Christian era, scholars-bibliographers of the Alexandrian libraries designed theories for organising books and for charting the human intellectual enterprise through indexes, lists of subjects, etc.

The monumental works of the sixteenth-century Swiss botanist and bibliographer Konrad von Gessner marked the beginning of modern bibliography. At the time of his demise in 1565, von Gesner had compiled a bibliography of existing documents in Greek, Latin and Hebrew.

The paternity of the term 'library science' (*Bibliothekswissenshaft*) is generally attributed to the German librarian Martin Schrettinger in the late nineteenth century. Buckland (2017a, p. 50) recalled the context in which Schrettinger coined the term:

> Around the end of the seventeenth century many monasteries in Europe were closed and their libraries confiscated. In Bavaria 200 monastic libraries were sent to Munich to be added to the royal library. The librarians were unable to cope with this flood of material until librarian Martin Schrettinger (1772–1851) understood that technical systems were needed to enable readers (as well as librarians) to

> find what they needed by themselves quickly and easily. It was for the technical guidelines that he coined the phrase *Bibliothek-Wissenschaft* (library science) in his textbook. (Schrettinger, 1808, p. 11)

Schrettinger devised a classification system for organising books on shelves according to topics and authors and provided shelf lists and a subject catalogue. Also the first journal in the field seems to be German: *Journal of Library Science, Manuscript Information and Older Literature* [*Serapeum: Zeitschrift für Bibliothekswissenshaft, Handschriftenkunde und ältere Litteratur*] was published in Leipzig by T.O. Weigel in the period 1840–1870 (Hjørland, 2017).

The late nineteenth century marked the emergence of systematic efforts to record published works and organise them in a systematic manner in libraries. In 1876, Melvil Dewey published the Decimal Classification (DC) in the United States enabling libraries to index and organise books. In Europe, Paul Otlet and Henri Lafontaine embarked on a monumental task to make the DC more powerful; this resulted in the Universal Decimal Classification (UDC) published in 1905. Schrettinger died two decades before Melvil Dewey published his DC in the United States and more than half a century before Paul Otlet and Henri Lafontaine published its European counterpart, the Universal Decimal Classification in Belgium in 1905.

Otlet's efforts to devise a bibliographic classification system was driven primarily by a positivist assumption that it was possible to extract knowledge from documents much in the same manner that one can extract peas from their pods. Otlet was obsessed by the desire to shed human knowledge of the messiness of the human subjective minds as reflected in books:

> Otlet's primary concern was not the document or the text or the author. It was also not the user of the system and his or her needs or purposes. Otlet's concern was for the objective knowledge that was both contained in and hidden by documents. His view of knowledge was authoritarian, reductionist, positivist, simplistic - and optimistic! Documents are repetitious, confusingly expressed and filled with error as well as with what is factually true and, therefore, of use. But he betrays no doubt that what is factually true and likely to be useful can easily be identified. It is merely a question of institutionalizing certain processes for analyzing and organizing the content of documents. For him that aspect of the content of documents with which we must be concerned is facts. He speaks almost everywhere of facts. (Rayward, 1994, p. 77)

Rayward (1997) observed that although Otlet and his colleagues did not use the term information science (IS), the "key concepts for information science as we now understand this field of study and research – and the technical systems and professional activities in which it is anchored – were implicit in and operationalized by what was created within the International Institute of Bibliography in 1895 and the decades that followed". Indeed, Otlet used 'documentation' and 'document' in a broader sense than the narrow one in which it was used by the profession of 'documentalists' in the post-Otlet period (Rayward, 1997, p. 290).

Otlet and La Fontaine's numerous international institutions – the *Institut International de Bibliographie* (IIB) in 1895 which became the *Institut International de Documentation* in 1931 and later the *Fédération Internationale de Documentation* (FID) in 1937, and his 1934 '*Treatise on documentation*' [*Traité de documentation*] – laid the theoretical and empirical foundations of what would become the field of IS as we now know it.

However, information and information science have remained fuzzy concepts with no clear meanings or boundaries as decades of debates have shown (Bawden & Robinson, 2012; Buckland, 1991; Capurro & Hjørland, 2003; Ibekwe-SanJuan, 2012d):

> One of the most serious problems confronting the historian of information science is knowing what it is that he or she is studying. Is information a process or a product? Is it text or document, the content of verbal communication, an expression of meaning, a statistical phenomenon of signal transmission, the processes of symbol representation and manipulation by electronic machines, biophysical activity of the brain, a matter of genetic or biochemical structures and processes? (Rayward, 1996, p. 3)

For the purposes of this book, we simply assume a practical definition of IS as that field of scholarly enquiry and teaching concerned with how people produce, organise, manage, record, use, disseminate and search for information for various uses. This rather pragmatic and processual definition purposely ignores the epistemological underpinnings of IS which we will not delve into here, more so as many publications have been devoted to these philosophical debates (Bates, 1999; Buckland, 2012; Cornelius, 2002; Furner, 2010; Hjørland, 1998; Machlup & Mansfield, 1983).

Several monographs and articles have also given historic accounts of how LIS emerged in different parts of the world, in particular in the United States (USA) and in the United Kingdom (UK) (Bawden & Robinson, 2012; Buckland & Liu, 1998; Rayward, 1983; Shera & Cleveland, 1985). Although some histories have been written about the emergence of LIS in non-anglophone European countries, they were published in languages other than English which make them difficult to access for an international audience.

Also, to the best of our knowledge, previous books on the history and foundations of LIS have focused on single countries. Cross-national studies of LIS are rare, especially concerning non-anglophone countries. Buckland and Liu (1998) noted this lack of cross-national intellectual histories of LIS.

The current book is an attempt to remedy this lack by providing the first of a series in cross-national histories of the emergence of LIS in non-anglophone European countries. It aims to highlight the characteristic features of LIS development despite apparent linguistic and terminological differences and the network of influences amongst European pioneers of LIS.

Our historical ambition is quite modest in that we are concerned only with the emergence of LIS as a higher education field of learning and enquiry, thus from the first quarter of the nineteenth century till the last quarter of the twentieth century. It is hoped that this tentative exploration into a comparative history of LIS in some non-anglophone European countries will inspire others to undertake more profound and complete histories of LIS in countries not dealt with here.

I.1. A Word on the Data Gathering Methodology

The choice of the European countries covered in this book was guided by pragmatic concerns, namely the availability of LIS scholars to interview, of funds to travel to these countries in order to conduct the ground research and, finally, of documentary materials in English or French, the two languages the author understands.

This resulted in seven countries that fall roughly into two geographic and cultural zones (apart for Yugoslavia which is Slavic):

(1) Latin Europe (France, Portugal and Spain) and
(2) Northern Europe (Denmark, Norway and Sweden).

This dependency on the availability of materials in English or French entailed an uneven treatment of these countries. France is treated in more depth since the author of this book resides and works in that country. She has also published a prior book in French (Ibekwe-SanJuan, 2012d) on the historical and theoretical foundations of IS in France in comparison with anglophone countries, mainly the US and the UK.

Thirteen people participated substantially in providing interviews for this research. Two people were interviewed separately in Zagreb for the Yugoslavian LIS, an emeritus LIS Professor and an administrative assistant who worked with the Croatian pioneer Božo Težak. The interview in Denmark involved an emeritus Professor of LIS whose extensive knowledge of the Nordic LIS landscape enabled him to cover the two other Nordic countries, Norway and Sweden. Three group interviews were carried out in Spain: four LIS scholars were interviewed at the University of Barcelona, another four were interviewed at Complutense University in Madrid, and two scholars were interviewed at Carlos III University in Madrid. A short biography of the interviewees is given in Annex I.

Oral histories are one of the established methods for reconstructing the history of disciplines:

> Oral history can be used in the historiography of science as an 'elite interview.' (...) 'Elite interviews can be very useful in eliciting otherwise unrecorded information and new perspectives (de Chadarevian, 1997, p. 53).' [...] To put it simply, oral history collects personal memories and personal commentaries of historical significance through recorded interviews. (Ritchie, 2003, cited in Hauk & Stock, 2012, p. 152)

To qualify as oral history, it is necessary to record the interviews such that the recordings can be archived and accessed for future historical research (Ritchie, 2003 cited in Hauk & Stock, 2012, 154). While recorded oral histories are more appropriate in the case of a historiography or archival project where the exact phrasing and expressions of the interviewee need to be preserved to serve as primary source material for other historical research, they were not appropriate in a book project where the exactitude of factual information and of ideas supersedes the need to record the exact utterances of the interviewee. Since our interviews of European pioneers and scholars were not recorded, it will be more appropriate to qualify our data gathering approach as a mixed qualitative approach combining semi-structured interviews with written documentary sources.

Questions were sent to all the interviewees in advance in order to give them time to recollect facts and gather materials for the interviews. This was a precautionary measure because of the transient aspect of human memory. Indeed, during the interviews, certain interviewees could not recollect exact dates nor the precise names of institutions and of pioneering figures and resorted to a Google search to verify some of the facts. Also sending our questions some weeks ahead of the interviews placed less tension on the interviewees who did not have to worry overly much about the conciseness and precision of their utterances, having been assured that all their utterances will be verified against published sources. Interviewees were thus placed in a more relaxed atmosphere and could focus more on narrating what they knew of events, people and places. Also, as language was a barrier for interviewees who were not proficient in the English language (this applied mainly to the interviews in Spain), sending the questions some weeks ahead of the interview enabled them to prepare their answers in English.

The duration of the interviews was not constrained but on the average they lasted between two to three hours, except for Denmark where the interview was spread over two days. Notes were taken during the interviews, edited and then sent back to the interviewees a few days later for correction and approval.

I.2. Caveats

Anyone who has attempted to write a history of ideas, of events and of anything would have felt the veracity of the following observation:

> History provides us with a way to think about the present and the future. Because we can never know it directly, it is actually constituted and reconstituted by what we bring to it from our ever-changing presents. It also offers the opportunity from the ever-changing perspective of the present to go back to reassess what seems to have happened, what seems to have been the case in the past and how it has influenced the present. It is this dialectical process that keeps history as a discipline always unfinished and alive. (Rayward, 2004, p. 671)

Also, as we have pointed out above, language proved to be an insurmountable barrier in the case of the Spanish LIS history where a significant amount of published works in Spanish could not be translated into English due to practical constraints. Hence, the account of the Spanish origins of LIS was based mostly on the information gathered during the interviews. Although we sought to cross check this information against published works, some aspects of the history will remain unverifiable because they are based on the subjective perceptions of the interviewees who knew some of the LIS pioneers and institutions in Spain. For the same reasons of language barrier, we are only able to offer a cursory history of Portuguese LIS based on a summary of pioneers made available to us in English. Again, these lines from Buckland (1999a, p. 9) cited in Hauk and Stock (2012, p. 152) serve as a timely reminder of the fallibility of oral histories:

> The whole process is accidental: who survived, what they knew, what they recalled, what they imagined, what they chose to relate, how they choose to express it and of course whether anybody bothered to record them.

I.3. Perimeter

Our history of the emergence of LIS in the seven European countries led us to constantly navigate between four types of entities situated at different levels:

(1) places where recorded artefacts of information and knowledge are conserved (documentation centres, archives, libraries and more rarely museums);
(2) educational and professional institutions (schools and universities) and associations (both professional and governmental) in charge of organising and regulating LIS education, profession and industries in these countries;
(3) influential figures (pioneers of LIS, administrators, political figures, etc.); and
(4) concepts (terminology) used to designate ideas, places and institutions that shaped the LIS in the countries studied.

These four entities were continuously interacting and changing over time in the seven countries. This can convey the impression of fundamental differences, but a deeper scrutiny tended to show that while the concepts, the terminology and the professional institutions varied from country to country, the four basic entities involved shared similar features across the seven countries studied. Hence, our challenge was threefold:

(1) to map out the particular characteristics of LIS development in each country;
(2) to identify common and distinctive features despite the apparent differences in terminology; and
(3) to highlight the network of external influences brought to bear on European pioneers which in turn determined the conception of LIS in a given country.

The 'histories' we have reconstituted in the following pages are neither complete nor irrefutable. However, we have reasons to believe that the common and distinctive features we uncovered on how LIS emerged in the seven countries are indicative of the trends in the other European countries between the early nineteenth and the last quarter of the twentieth century.

One of the commonalities that came to light is the observation that the ideas, perspectives and paradigms underlying the first LIS education programmes in many European countries at the beginning of the nineteenth century were influenced by the French perspective which itself was rooted in the French Revolution of 1789. The French model of separation between vocational schools outside of academia, in charge of training archivists, librarians and documentalists, of which a typical example is the *École Nationale des Chartes* established in Paris in 1821, was adopted in several European countries (Ribeiro, 2007, p. 115). We observed in all the countries we studied that two educational systems for LIS co-existed in many European countries well into the first half of the twentieth century. Special vocational schools trained professionals in areas perceived as crafts or vocations such as nurses and teachers. Universities were considered as the academic seats of learning and were in charge of training doctors, scientists and scholars. The management of libraries was considered to be a vocational occupation; therefore, the first training programmes were handled by special schools outside of universities.

Figure I.1 gives an overview of the earliest formal education in librarianship which was outside of universities in the seven countries we studied.

Around the 1970s, these vocational library schools began collaborating with universities. The move towards universities as the seat of education in LIS was accompanied by a broadening of focus from purely material and technical aspects of library management to more theoretical and interdisciplinary content as well as an alignment with university degrees:

> By the 1970s schools of librarianship worldwide were gradually broadening their interests and redefining their scope to include librarian-like activities outside of libraries as well as inside. There were several reasons: there was a surplus of librarians and a shortage of information professionals adequately equipped for similar work in other contexts; research universities preferred graduate programs to address a type of expertise rather than a type of institution; to the extent that problems in other contexts were similar, theories and technical solutions could be extended, challenged, and made more robust; and so on. Yet it was not only a change in scope, there was also a strong desire to evolve a new and different way to view and understand the field itself. (Buckland & Lund, 2013, p. 303)

The background of teachers was also broadened to include scholars from science and engineering who embraced the emerging information technology as a way to solve problems of indexing and document retrieval. As a result, the field slowly acquired scientific status and gained recognition as an academic discipline. As separate library schools began to lose their monopoly on library training, they were either absorbed into universities or relied on the latter for delivering LIS diplomas after accreditation by the designated regulating bodies. Figure I.2 provides an overview of this evolution towards universities.

In the following chapters, unless stated otherwise, translations of original texts from other languages into English are provided by the author.

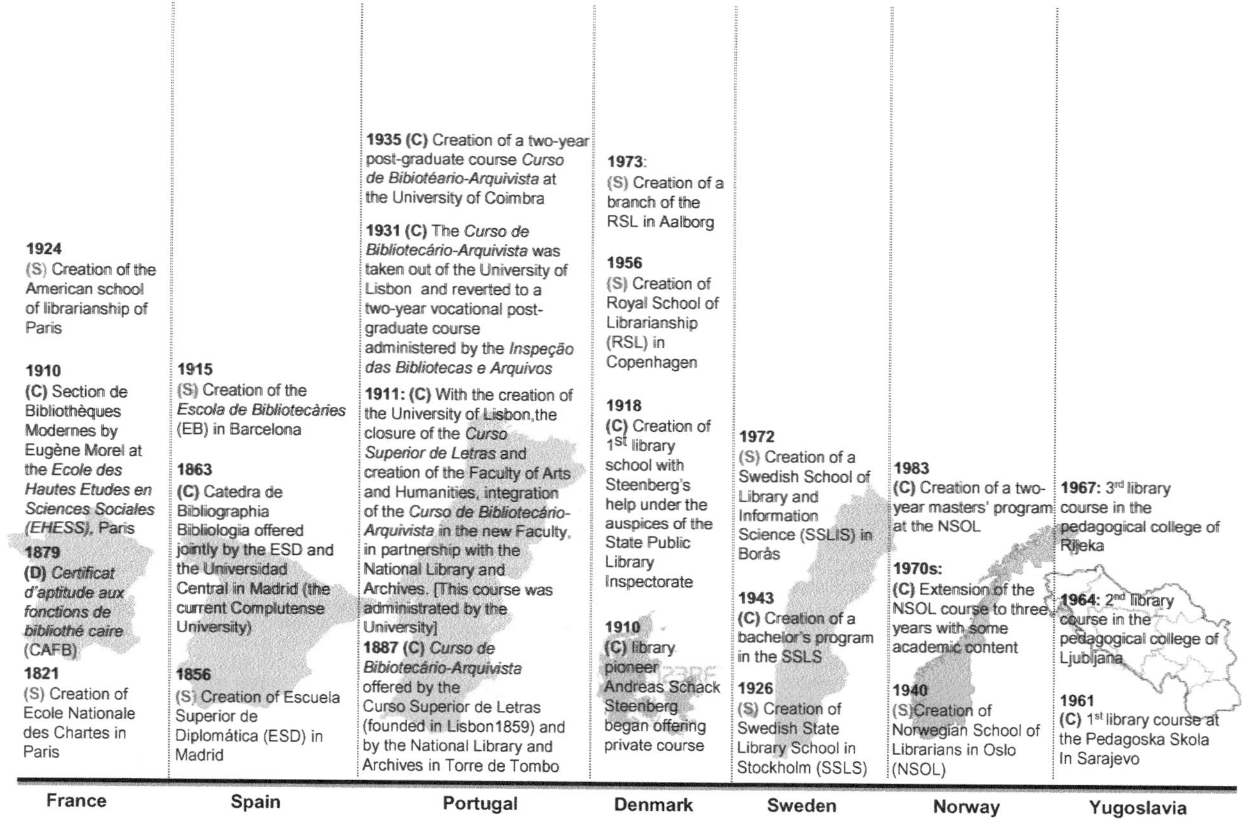

Figure I.1: Timeline of the earliest training programmes in Librarianship in special schools outside of universities. *Key*: C = Course; D = Diploma/Certificate; S = School; I = Infrastructure.

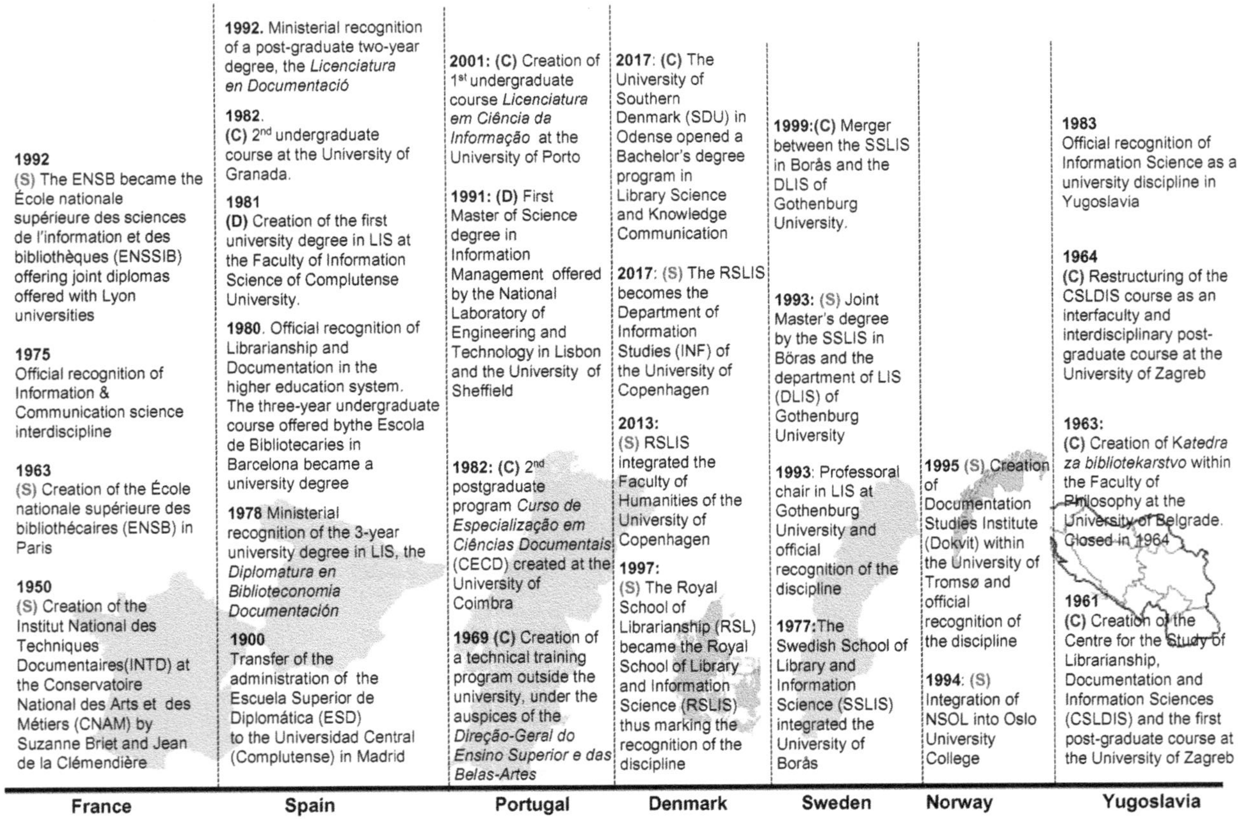

Figure I.2: The integration of library training into universities towards documentation, information science and official recognition as an academic discipline. *Key*: C = Course; D = Diploma/Certificate; S = School; I = Infrastructure.

Chapter 1

Emergence of LIS in France: A Tale of Academic Cohabitation and of Memory Loss

This chapter recalls the earliest formal training in librarianship (Section 1.1), the role Eugène Morel played in the modernisation of libraries in France (Section 1.2) and the role of Suzanne Briet in creating the first documentation programme which foreshadowed the emergence of Information Science (IS) in France (Section 1.3). The winding road from documentation to IS is discussed in Section 1.4, and the official recognition of IS within the interdiscipline called Information and Communication Sciences (ICS) in the French higher education system and the first university programmes are discussed in Section 1.5.

1.1. The Earliest Training Programmes in Librarianship

The emergence of Library and Information Science (LIS) in France followed a centripetal[1] model. Until the second-half of the twentieth century, all the training courses in librarianship were concentrated in Parisian institutions. This is not altogether surprising when one considers that France is a highly centralised state where professional and public activities are regulated by the central government. Until the second-half of the nineteenth century, it was possible to become a librarian in France without necessarily having a degree in librarianship (Ducolomb, 2012[2]). A similar situation was observed in other European countries at the same period.

[1]Moving or tending to move towards a centre.

[2]Our account of Eugène Morel's contribution to the modernisation of French libraries is based on Ducolomb's (2012) article retrieved from http://bbf.enssib.fr/consulter/bbf-2012-01-0035-006. Consulted on 6th January 2018.

The first French institution to offer a formal library education was the *Ecole de Chartes*, founded in Paris in 1821 by order of Louis XVIII. The aim of the school was to train historians, archivists and palaeographers to restore France's history to its former glory by studying old manuscripts and charters confiscated during the French Revolution (1789–1799). Hence, the *Ecole de Chartes* was not designed as a library school originally, but as time went on, it began to offer library courses. As one of France's special elite academic institutions [*Grande écoles*], entrance into the *Ecole de Chartes* is by examination, and selected pupils become trainees of the State who are paid to study.

In 1822, a national body that regulates libraries, the "General Inspection of libraries" [*Inspection générale des bibliothèques*] was founded. In 1879, the first professional certificate in librarianship, the "Certificate of Aptitude to the functions of a librarian" [*Certificat d'aptitude aux fonctions de bibliothécaire* (CAFB)] was created. Gradually, other professional exams were put in place. By the beginning of the twentieth century, the necessity to broaden the scope of the training offered by the *Ecole des Chartes* became apparent.

The American Committee for Devastated Regions of France [*Comité Américain pour les Régions Dévastées* (CARD)] was created in 1917 to spearhead the reconstruction of damaged infrastructures and services during the First World War. In 1924, the Paris Library School was created by the American Library Association (ALA) and the Library of Congress[3] as part of the US's post-war efforts to rebuild devastated regions. This school, which functioned until 1929, played a crucial role in the reconstruction and the modernisation of French libraries. Whereas the profession had been male dominated, the ALA's Paris Library School trained many women librarians some of whom were sent to the US to learn the current techniques for managing libraries. This school recruited French librarians who were enthusiasts of the modernisation efforts and of the American free open libraries approach. As Fayet-Scribe (1997, p. 791) acknowledged, "public librarianship [in France] owes much to the American model".

With the creation of the "technical diploma of librarian" [*Diplôme technique de bibliothécaire* (DTB)] in 1932, the *Ecole de Chartes* began to offer a specific training programme for librarians which consisted of only two courses: library management [*bibliothéconomie*] and bibliography. In 1950, the DTB was replaced by the "Higher Diploma of

[3]A historic account of the American Library created in Paris in 1920 which led to the creation of this school by the American Library Association (ALA) and the Library of Congress can be found here: https://americanlibraryinparis.org/history/.

Librarian" [*Diplôme Supérieur de Bibliothécaire* (DSB)]. From the 1950s, the CAFB diploma underwent several revisions until it was finally withdrawn in 1992. The same year (1992), the DSB gave way to the "Diploma of conservator of libraries" [*Diplôme de conservateur des bibliothèques* (DCB)] which is the current higher degree for librarians working in the public sector. In 1963, a dedicated library school, the "Higher School of Librarianship" [*Ecole Nationale Supérieure de Bibliothécaires* (ENSB)] was also created in Paris. It was modelled after the ALA's Paris Library School. In 1974, the ENSB was transferred to Villeurbanne in the Lyon region,[4] thus constituting the first "delocalisation" of a LIS institution from Paris to the Province.

The year 1992 was a landmark year for LIS training in France. It not only marked the transformation of the ENSB into the "Higher School of Information and Librarianship" [*Ecole Nationale Supérieure des Sciences de l'Information et des Bibliothèques* (ENSSIB)] which conferred it a higher education status, it also signalled the merging of librarianship with the emerging Information Science. The ENSSIB has close ties with the adjacent big science university in Lyon (Université Claude Bernard), but it has its own Director and administrative council. The year 1992 was also synonymous with national reforms that culminated in the homogenisation of the different professions making up the public library sector (*conservateurs, bibliothécaires, magasiniers, …*).

Figure 1.1 shows the earliest formal training concentrated in Paris before other centres appeared in the 1970s elsewhere in France.

With the notable exception of an article in English by Sylvie Fayet-Scribe (1997, 2000), most of what we know about the history of LIS in France is written in French. Our aim is not to rewrite this history in English. Rather, we have chosen to focus on two pioneering figures: Eugène Morel and Suzanne Briet.

[4]More details about the first educational programmes and diplomas in librarianship can be found in this account written by the French Ministry for Public Service: "LES MÉTIERS DE LA FONCTION PUBLIQUE. Bibliothécaire: l'information à livre ouvert", Retrieved from https://www.fonction-publique.gouv.fr/archives/home20051107/fp/metiers/bibliothecaire.htm. Accessed on December 14, 2017.

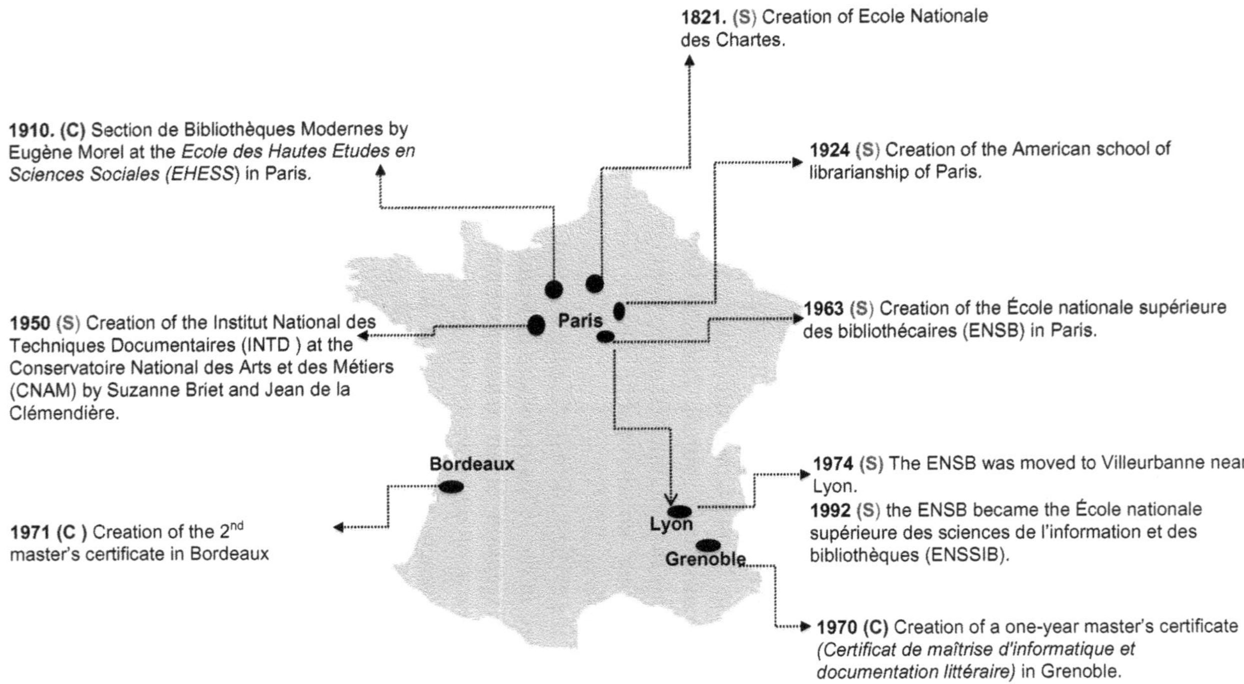

Figure 1.1: The Earliest Institutions for LIS Education in France. *Key*: C = Course; D = Diploma/Certificate; S = School; I = Infrastructure.

1.2. Eugène Morel, a Pioneer of Library Modernisation

After a degree obtained from the University of Paris, Eugène Morel (1869–1934) embraced a career in librarianship while being a writer and a literary critic. He began working at the Bibliothèque Nationale de France (BNF) in 1892 as an assistant librarian. At the beginning of the twentieth century, Morel published two important books on the necessity to modernise the training for librarians and consequently improve their services to the public: "Libraries. Treatise on the development of public libraries and of bookshop in the two worlds"[5] [*Bibliothèques. Essais sur le développement des bibliothèques publiques et de la librairie dans les deux mondes*] in 1908 and "The public library"[6] [*La Librairie Publique*] in 1910 in which he detailed his project to modernise the training of librarians in order to prepare them for the future. In 1906, Morel also co-founded the French association of librarians [*Association des bibliothécaires français* (ABF)], of which he became president in 1918. The ABF is still the main association of French librarians.

Morel was a fierce critic of the inadequacy of the library courses delivered by the *Ecole des Chartes*. He considered it as being heavily oriented towards medieval history and palaeography, to the detriment of the skills and knowledge needed to work in public libraries. In *La Librairie Publique*, he deplored the fact that historians and literary scholars have always had a stranglehold on the education and profession of librarian and called for an opening to other specialties and competences:

> Two specialties have until now almost exclusively disputed libraries: literature and history. By opening general libraries up to men, to others, we must seek to vary a little the competencies and the influences.[7]

The consequence was that librarians, trained in separate library schools (Ecole de Chartes, ENSB) were not well prepared either to serve the general public or to adjust to technological developments that have greatly impacted the field.

---

[5]Eugène Morel, *Bibliothèques. Essais sur le développement des bibliothèques publiques et de la librairie dans les deux mondes*, Paris, Mercure de France, 1908. Retrieved from www.enssib.fr/bibliotheque-numerique/document-48751.

[6]Eugène Morel, *La Librairie publique*, Paris, Armand Colin, 1910. Accessible online www.enssib.fr/bibliotheque-numerique/document-48832.

[7]Our translation from the French original "*Deux spécialités se sont jusqu'ici presque exclusivement disputé les bibliothèques : la littérature et l'histoire. En ouvrant aux hommes – aux autres – des bibliothèques générales, il faudrait surtout chercher à varier un peu les compétences et les influences*" - *Morel E., La Librairie publique, op. cit.*, pp. 288–289.

Ducolomb (2012) observed that the situation decried by Morel at the beginning of the twentieth century still obtains today. The "preserver of library"[8] [*conservateur de bibliothèque*] is protected by a special examination prepared by special schools and this exam mostly requires that they are well versed in "general culture" [*culture générale*] which is mostly history and literature. Few certified librarians who work in science libraries have a deep knowledge of the subject matters in which their host institutions are specialised.

A keen admirer of the American library system, Morel took several trips to the UK and the US which shaped his conception of the modern librarian. From those trips, he became an advocate of the anglophone model of the free public library where access was open to anyone and libraries were publicly funded. Benoït (2008) gave an account of Morel's involvement with the ALA's Paris Library School effort to modernise the profession in France. Morel wanted the librarian in public libraries to constantly seek to engage the public by providing all sorts of information, not only bibliographic information from library catalogues but also useful information from newspapers and other resources. He saw libraries and librarians not as "preservers" [*conservateurs*] of printed materials but as information hubs that put their patrons at the centre of their activities. A historic account by Hassenforder (1967) explained why public libraries in France were so different from those in the UK and the US.

In 1910, with the support of the newly founded ABF, Morel set up a training course called "Section of modern libraries" [*Section de Bibliothèques Modernes*] which was housed within the "Higher School of Social Sciences" [*Ecole des Hautes Etudes en Sciences Sociales*], another of France's *Grande écoles* [elite higher education institutions]. The organisation of this training programme was based on the model established by the British Library Association. This course was run as a cycle of lectures spread over three years at the end of which the trainees had to gain practical experience during a three-month internship in France or abroad, overseen by the ABF. To obtain their diploma (*brevet*), trainees thus had to attend lectures, acquire practical experience through an internship and write a *mémoire* at the end of the training period.

Speakers invited to Morel's librarianship lectures came not only from the library profession but also from other areas of life. They included publishers, printers, journalists or even bankers with personal library collections. In December 1910, the founding fathers of documentation and of the

---

[8]We have given a literal translation to emphasize the signification of the French word "conservateur" which comes from the verb "conserver", meaning to preserve well. The closest actual meaning of "conservateur de bibliothèque" would be "library curator"; however, curator is usually used in association with museums in English, rarely with libraries.

International Institute of Bibliography (IIB), Paul Otlet and Henri La Fontaine were invited to present, in front of the students enrolled in Morel's course, the latest techniques of the profession where the future librarians would deliver not only documents to scientists but also *dossiers* that are constantly updated on specific topics. This foreshadowed the emergence of "selective dissemination of information" [*diffusion sélective d'Information* (DSI)] in the 1980s and of science and technology monitoring as a specialty in the LIS field. More importantly, the presence of Otlet and La Fontaine in these lectures brought international recognition to Morel's endeavours. It also gave him much needed support in the face of opposition from the more traditional *Ecole de Chartes* and signalled that documentation and information were now part of the training for librarians.

In 1911, Morel introduced the Dewey's DC system to France by implementing it at the Levallois-Perret Library in Paris.

With the outbreak of the First World War in 1914, Morel's *Section des bibliothèques modernes* lecture series was closed down.

Other training programmes emerged after the First World War. The ALA's Paris Library School where Morel taught until 1926, finally closed in 1929, but its short existence (1924–1929) had a significant impact on the library profession in France. From the 1930s, other *ad hoc* training programmes for librarians emerged: there was a training programme within the Forney library from 1930 to 1936 and another one in the school of librarians at the Catholic Institute of Paris.

Eugène Morel can be seen as a precursor of the library 2.0 movement where the emphasis has shifted from the librarian's focus on knowledge organisation (cataloguing, classification) and preservation of documents to the social functions of the library, open to the public for various uses. Morel was not only a precursor of the modernisation of library training, he was also a forerunner in the emergence of training programmes for information professions in France.

1.3. The Heritage of Suzanne Briet, France's *Madame Documentation*

While Morel was the leading figure in the modernisation of libraries and of library education in France, Briet was certainly the leading figure in the transition from librarianship to documentation and subsequently to Information Science (IS). Bruno Delmas, former Director of the National Institute for Documentary Techniques [*Institut National des Techniques Documentaires* (INTD)], an institution that Briet co-founded, gave two interesting accounts of the history of documentation in France and of

Briet's role in it. As a concept, documentation emerged in the French vocabulary in the mid-eighteenth century, at a time when "France was one of the most advanced countries in Europe for scientific research and technical studies. Since the reign of Louis XIV (1638–1715) but especially in the second-half of the 18th century, a series of schools and academies were created where special libraries had been constituted" (Eric de Grolier cited in Delmas, 1992, p. 179). Concepts designating this new specialty appeared in this period: the verb "to document" [*documenter*] which meant "to provide documents" appeared in 1769; the noun "documentation" which meant the "action of seeking for documents to support a study" made its appearance in 1870, a century after the verb; the adjective "documentary" [*documentaire*] which meant "that which has the character of a document or that which is based on documents" appeared in 1876 (coincidentally, the same year Dewey published his DC).

1.3.1. The Institutional and Historical Context Preceding Briet's Appearance

Now that the terminological stage has been set, institutions and humans began to inhabit the star constituted by documentation in the librarianship universe. From 1880 onwards, documentation and bibliography emerged as distinct specialties from librarianship. For a while, other European countries took the lead in inventing innovative documentary techniques and institutions (Delmas, 1992, p. 180). In Belgium, Paul Otlet and Henri La Fontaine were very actively creating bibliographic institutions. In France, Général Hyppolite Sebert, a graduate of the Polytechnic school [*Ecole Polytechnique*] who also had a military career before becoming interested in the advancement of science, became the vice president of the IIB which he helped found with Otlet and La Fontaine. According to Fayet-Scribe (1997, pp. 783–784), "Otlet and Sebert enjoyed a close, long-time relationship, as did Otlet with his colleague Henri La Fontaine, as shown by their correspondence from 1898 to 1921 {*Dossier N260:* Letters from General Sebert to IIB, and letters from IIB to General Sebert, Mons, archives du Mundaneum)." The IIB gave a new impetus to bibliography and documentation in France. In 1902, the "Bibliographic Bureau of France" [*Bureau Bibliographique de France*] BBF, the French arm of the IIB was created in Paris and managed by Sebert. Its objective was to compile a *repertoire* of printed scientific and technical documentation in France. The *Bureau* gathered librarians interested in the new documentation techniques and bibliography. Amongst its most well-known members were Eugène Morel, the French "revolutionary librarian *extraordinaire*", Charles Sustrac and Armand Boutillier de Retail. Their primary goal was the dissemination and implementation of Otlet's Universal Decimal Classification (UDC); hence, they were known as the "decimalists". After Sebert's retirement, Charles Bayle and Eugene Morel played prominent roles in the leadership of the BBF. After Morel's retirement, Georgette and Eric de Grolier, husband and wife became involved in the BBF.

In 1901, a National Office for Intellectual Property [*Office National de la Propriété intellectuelle*] was created within the "National Conservatory of Arts and Occupations" [*Conservatoire National des Arts et des Métiers* (CNAM)] in Paris to manage the documentary activities involved in searching for priority and the dissemination of patents of invention. This was followed by the creation of specialised information and documentation centres in different sectors of the industry. Gradually, other sectors of activity created their own information and documentation centres. Scientific and technical documentation centres began to appear in universities and other institutions of learning. In the first decade of the twentieth century and under Paul Otlet's influence, documentation appeared as the "new order" for organising, searching and retrieving recorded information (Delmas, 1992, pp 180–181). The outbreak of the First World War slowed down the efforts to create specialised bibliographies and documentation centres, but they picked up again at the end of the war.

1.3.2. Suzanne Briet: Career at the Bibliothèque Nationale de France

Sadly, Briet's contribution to the emergence of IS had been neglected by French IS scholars until her "rediscovery" in the 1990s by North American scholars (Buckland, 2005; Day & Martinet, 2006a, 2006b; Day, 2007; Maack, 2004).[9] The synthesis we offer here of Briet's contribution to the emergence of documentation and of IS worldwide owes much to these earlier accounts in English, augmented by our own reading of Briet's 1951 manifesto in French "*Qu'est-ce que la documentation*" [What is documentation?].

Suzanne Briet (1894–1989) majored in History and in English and worked as a high school teacher before training to become a librarian in the early twentieth century. She was one of the few women to have qualified as a librarian in 1924, at a time when the profession was dominated by the masculine gender in France, unlike in America where "By 1910, 78 percent of library workers in the United States were women" (Fasick, 1986, p. 608).

[9]A compendium of articles on Briet written by American LIS scholars is hosted by Michael Buckland at http://people.ischool.berkeley.edu/~buckland/briet.html and by Ronald Day at http://ella.slis.indiana.edu/~roday/briet.htm.

Briet's 30 years career at the French National Library [*Bibliothèque Nationale de France* (BNF)] spanned the period before and after the Second World War. Under the guardianship of Roland Marcel who became the Director General of the BNF in 1923, an Office of documentation was created within the library to handle requests for information and for document delivery. The rationale for this new service was according to Briet herself that librarians at the BNF, who were already "overburdened" could not effectively respond to written requests for information because such work interrupted the normal flow of activity. Thus, Briet was put in charge of the first referral service [*centre d'orientation*] inside the BNF to alleviate the burden of the librarians who must not be distracted from their more noble and traditional cataloguing tasks. In June 1928, she was given the responsibility to bring together all such requests (Maack, 2004, p. 725). This new documentation service was in charge of responding to requests for scientific literature both from French and foreign scholars. The "Room of catalogues and of bibliography" [*Salle des catalogues et de bibliographie*] created at the BNF in 1931 became Briet's domain over which she presided for over two decades (Delmas, 1992, p. 181). By the end of the 1920s, Briet made reference works more accessible to patrons, whereas these would normally be stacked in the storerooms, out of reach of library users. This was in itself a revolution in the traditional library, usually focused on preserving books with little attention paid to other types of documents.

Briet benefitted from the invaluable support of the two Directors of BNF under whom she served, Roland Marcel and Julien Cain. Julien Cain (1887–1974) deserves a special mention in that he played an important role in modernising the BNF in the first-half of the twentieth century. He majored in history before being conscripted to fight during First World War. Having sustained significant wounds and undergone two operations, Cain was declared unfit for military service. In 1917, he was assigned to the foreign documentation service of the Ministries of Foreign Affairs and of Defense. In 1930, he was nominated as Director General of the BNF with the explicit goal to modernise this very conservative and fossilised institution, a task he undertook with vigour and had mostly completed by 1935. During Second World War, he was interned in a German concentration camp at Buchenwald. Having escaped death, Cain was restored to his former position as Director of the BNF after 1945 and pursued his library modernisation *oeuvre* with renewed vigour. Being progressive, he held ecumenical views of librarianship, documentation and information services. He hired the first female professional librarians at the BNF, one of whom was Briet. A versatile cultural activist on several fronts (politics, literature and information services), he was actively engaged with several library associations both in France and abroad. He launched the major French professional library journal *Bulletin des Bibliothèques de France* (BBF) in 1956.[10] He was at a time vice president of the International Federation of Library Associations (IFLA). In 1964, he retired after having been at the helm of the BNF for close to 30 years (1930–1940; 1945–1964).

1.3.3. Suzanne Briet: The Creation of the First Documentation Association and the INTD Years

Like Julien Cain, Briet was an active member of national and international library associations. With the support of Julien Cain and of Jean Gérard, another advocate of documentation techniques in the chemical field, Briet co-founded the first French Union of documentation organisations, *Union Française des Organismes de Documentation* (UFOD)] in 1931, six years before the American Documentation Institute (ADI[11]) was created in the US but seven years after the Association of Special Libraries and Information Bureaux (Aslib) was created in Britain in 1924. From 1932, the UFOD began publishing a monthly bulletin called *La Documentation en France* in which Briet wrote a column entitled "New bibliographies" [*Bibliographies nouvelles*] (Delmas, 1992, p. 187).

From 1932, the UFOD set up a committee to work on a specific educational programme to train future documentalists. Mandated by the UFOD to lead the work of this committee, in 1934, Briet presented the project of a school of documentation to the UFOD, but the outbreak of Second World War and the subsequent economic depression it brought in its wake hindered the realisation of this project. In January of 1945, after the liberation of France and with the support of UFOD, Briet and Jean de la Clémendière opened the first course in technical documentation while the country was still in disarray and people were suffering from severe hardships. Nevertheless, this new course attracted 70 pupils most of whom worked in either government agencies or private organisations (Delmas, 1993, pp. 219–220). This documentation course was officially recognised in December 1950 by a ministerial decree, and it became known as the "National Institute for Documentary Techniques" [*Institut National*

[10]Interested readers can find out more about this very industrious library, literary as well as political activist on his Wikipedia page: https://fr.wikipedia.org/wiki/Julien_Cain.

[11]ADI was renamed the *American Society for Information Science* (ASIS) in 1968, then *American Society for Information Science & Technology* (ASIST) in 2000. In 2012, in keeping with the internationalisation of its membership, an overwhelming proportion of the membership voted to drop the "American" in its name. It is now simply called the "ASsociation for Information Science & Technology" with the same acronym.

des Techniques Documentaires (INTD)] and was housed within the Conservatoire National des Arts et des Métiers (CNAM) in Paris. Briet was the Director of studies of the INTD between 1950 and 1954. The official text promulgating the creation of the INTD stipulated that it may carry out all research that enabled the improvement of documentation techniques and tools (Delmas, 1993, pp. 221–222). The INTD's programme was structured as a two-year course: the first year was designed as a general introduction to library and documentation techniques (acquisition, cataloguing, indexing, dissemination and reproduction of document) while the second year was designed as a specialisation year in documentation in a chosen subject area. The programme was fee paying and open only to French nationals below 30 years of age, although foreigners were allowed to attend classes freely, as auditors (Delmas, 1993, p. 218). Until the creation of the ENSB in 1963, the INTD was the only place in France where a formal accredited degree in documentation could be obtained. Its first graduates were thus pioneers of the profession. From 1970, the INTD began expanding the duration of its programmes to align them with an engineer or masters level. By the mid-1970s, it had dropped the first undergraduate-level technical programme in documentation in favour of advanced master's and engineer programmes. This reduced the intake of students to people who already had a first degree in a specific subject area or were embarking in continuing education. Like the ENSSIB in Villeurbanne, INTD's degree programmes were gradually aligned with a university-style education requiring full time students' attendance with the result that the INTD also became a "Higher School of Information Science" [*Ecole supérieure des Sciences de l'information*]. A more detailed description of the different diplomas offered at the INTD until the mid-1990s can be found in Delmas (1993). The INTD still exists today within the CNAM. Together with the ENSSIB, it is considered as one of the top places (*haut lieux*) of higher education in IS and in documentation. The INTD can be considered as the first *iSchoolish* kind of institution in France. This would make Briet the first Director of an iSchool. As the "founding director of studies of the educational program at the Institut National des Techniques de la Documentation in Paris in 1951", Suzanne Briet "has a claim to being the founding mother of the i-School movement" (Buckland 2017a).

1.3.4. Suzanne Briet: American Influence on French Documentation

Being fluent in English and having travelled to England during her childhood, Briet was able to take advantage of advances in librarianship in the UK and the US. A Fulbright grant enabled her to travel to the United States between 1951 and 1952 to learn how American libraries were being run. She observed that the new documentation centres she was setting up in France existed in the US and in the UK but under different names. As Buckland (1996, p. 74) wrote, "The *terms* "documentalist" and "documentation center" were not in use in the US, but the *roles* of documentalist and of documentation center were occupied early on by special librarians and special libraries even though no courses on documentation or even special library work were offered by US library schools until after the Second World War. Briet had the insight to recognise that this was more a difference in name and appearance than in substance (Briet, 1953, 1954; Buckland, 1996; Williams, 1997)."

The 1930s and 1950s appear to have been the golden years of documentation in France. This period coincided with the position of Paris as the centre of scientific, artistic and intellectual activities in Europe. Paris shared with Brussels the title of the capital of documentation. The first World Congress of Universal Documentation was organised in Paris in 1937 during the International Exhibition of Art and Technology in Modern Life. It was during this congress that the creation of the Fédération Internationale de Documentation (FID) was decided and endorsed in Oxford in 1938. Headquartered at The Hague, the FID took over some of the activities of Otlet's IIB, which by then had been renamed International Institute of Documentation [*Institut International de Documentation* (IID)] in 1931 (Delmas, 1992, pp. 185–187). In the 1940s, different laws were passed in France authorising the creation of information and documentation centres in public institutions and in government agencies. The scientific documentation centre of the French National Research Center [*Centre National de Recherche Scientifique* (CNRS)] was created in 1942 (Delmas, 1992, p. 191). The UFOD played a key role in organising the profession and in obtaining recognition from the national government.

1.3.5. The Lady and the Antelope.[12] Briet's Semiotic Conception of Document

Buckland (2018) observed that 1951 was "a triumphant year for Briet because in that one year, she published her famous manifesto *Qu'est-ce que la documentation*; her program of professional education for documentalists at the CNAM received official recognition in the shape of the INTD; she was vice president of the FID; and she met and talked with the Pope when she led the French delegation to the FID conference in Rome".

[12]This phrase is borrowed from the excellent biography of Suzanne Briet written by Mary Niles Maack (2004).

Qu'est-ce que la documentation[13] is an illustration of a "sleeping beauty" type of publication, that is, a piece of work barely acknowledged at the time of its publication but which is "re-discovered" decades later and which makes a significant contribution to the understanding of an important phenomenon in a scholarly field of study. The core contribution of Briet's manifesto lies in her broad conception of document and of documentation. She began by asking:

> Is a star in the sky a document? Is a pebble rolled around by a torrent a document? Is a living animal a document? No. But photographs and catalogues of stars, stones in a mineralogy museum, animals catalogued and exhibited in a zoo are.[14] (Briet, 1951)

She illustrated this broad conception of the document by walking the reader through the example of the different documentary manifestations that the discovery of a new species of an antelope in Africa can give rise to. Supposing that the living animal was brought back to Paris and kept in a botanical and zoological garden in Paris (*Jardin des Plantes*), it would have been "classified" according to the prevailing zoological classification scheme, thus contributing to the advancement of knowledge in the field of zoology. Once dead, it would be preserved and placed in a museum of natural history and loaned to other cultural institutions for exhibitions. Its discovery would give rise to multiple documentary productions on different media and for different sets of audience: press releases for the public, lectures and seminars by experts, articles in journals, illustrations and drawings in books, statues, paintings, photos, documentary films, books, encyclopaedia entries and radio programmes. All the above documentary productions related to this new specie would have been equally recorded in library catalogues as well as in publishers' catalogues. Each new publication on the antelope would be deposited at the BNF as warranted by the law and each in turn would receive a documentary analysis in library catalogues.

The question raised by Briet is therefore which of these instances of intellectual productions around this antelope is a document? Her answer is unequivocal: the living antelope from Africa captured and exhibited in a cage, a zoo or a botanical garden is "an initial document". All the other

---

[13]An electronic version of her book with Laurent Martinet's preface is accessible at à http://martinetl.free.fr/suzannebriet/questcequeladocumentation/.

[14]In the original text, Briet wrote: "Une étoile est-elle un document? Un galet roulé par un torrent est-il un document? Un animal vivant est-il un document? Non. Mais sont des documents les photographies et les catalogues d'étoiles, les pierres d'un musée de minéralogie, les animaux catalogués et exposés dans un Zoo."

intellectual productions in support of the proof of the existence of this antelope are "*secondary* or *derived* documents".[15]

A further distinction should however be made within Briet's secondary or derived documents to distinguish between original intellectual productions on the existence of the antelope, such as books, articles, recordings, and paintings which are *primary documents* and the documentary analysis performed on these *primary documents* by information professionals (documentalists, librarians or museum curators) when classifying them in libraries catalogues, producing bibliographic records, summaries and index terms, which constitute *secondary* documents (in documentary parlance). Hence, there is a hierarchy of documents here: the living captured antelope is an "initial document" according to Briet. Original intellectual productions about this antelope are *primary* documents while the products of documentary analysis of these primary documents (catalogues, index terms, summaries, bibliographic records, subject headings, etc.) are *secondary* documents.

Having said what "things" can be a document (almost anything), Briet then gave her definition of document as "any concrete or symbolic indexical sign [*indice*], preserved or recorded, for the purposes of representing, of reconstituting, or of proving a physical or intellectual phenomenon".[16] This definition surpasses in scope the traditional definitions of a document which tended to limit it to the printed material, often of a scientific, religious or professional nature. Briet's document can take any shape or form, thus she anticipated well ahead of time the impact of the digital in the "virtualisation" and fragmentation of documents. Through the antelope example and her definition, Briet demonstrated the fecundity and versatility of the concept of document, which despite many ominous predictions about its obsolescence due to the Internet and the digital age continues to be as relevant as ever.

One may wonder how Briet came by such a broad conception of a document from working in a national library which at the time was mostly concerned with preserving books. Briet's biographers have remarked that for a librarian, she was rather vague about her bibliographical inspirations. Several avenues have been explored as possible inspirations for her "antelope-as-document" example. One plausible explanation is that the BNF where Briet spent 30 years also had a collection of objects (coins, sculptures, medals, maps and personal souvenirs). Briet had suggested using the term "documentography" [*documentographie*] instead of

---

[15]In the original French text, Briet (1951, p. 8) wrote "*L'antilope cataloguée est un document initial et les autres documents sont des documents seconds ou dérivés.*"

[16]The English translation of Briet"s book by Day and Martinet (2006a, 2006b). In the original French text, Briet's definition of a document is "*tout **indice** concret ou symbolique, conservé ou enregistré, aux fins de représenter ou de prouver un phénomène physique ou intellectuel*".

bibliography to acknowledge the versatility of information bearing media, thus of documents. It also appears that this broad conception of a document was not uncommon in the European documentation tradition of the time around the institutions built by Otlet (IIB, IID and FID). This was long before Briet wrote her manifesto. However, the exact source of inspiration for her "antelope-as-document" example remains obscure.

Michael Buckland, who has been studying her writings (amongst other European LIS pioneers) observed that, like Paul Otlet, Briet's writings are very declarative and often omitted to make explicit her sources. He uncovered a second possible inspiration for the Briet's "antelope-as-document" example: Briet may have been inspired by a prior thesis written by one of her students, Robert Pagès, another French pioneer of documentation and of IS (see Section 1.4.1 "The beginning of mechanised documentation" hereafter):

> "[…] in 1946, Pagès became a student in the documentation program at CNAM founded by Jean de la Clémendière and Suzanne Briet and later designated the INTD. In 1947 he wrote a short thesis entitled "Transformations documentaires et milieu culturel". The next year it was published as an article in the leading journal of documentation, the *Revue de la documentation* published by F.I.D. The article has the same title "*Transformations documentaires et milieu culturel: (Essai de documentologie)*" [Documentary transformations and cultural context: (Essay on documentology)] [PAG 48]. In Pagès' article are the same ideas as those published three years later in Briet's manifesto, but, unlike Briet, Pagès provides a clear explanation. In brief, graphic documents, which describe or represent something else, are derived, or "secondary", documents. But specimens and other non-graphic objects only represent themselves (as "auto-documents") and so are primary documents. Pagès' examples include a gorilla in a cage rather than an antelope, but it is clearly the same idea. So it seems that the teacher (Briet) learned from the student (Pagès), but maybe not. Maybe these ideas came from Briet long before she wrote her manifesto or came from some other source. It is remarkable that although the antelope has become famous the gorilla appears to have remained unnoticed." (Buckland, 2018)
>
> There is scope for further research on this matter.

1.3.6. Briet's Documentalist as an Assistant to the Scientist and the Specialist

For Briet, the unit of documentary analysis is not far from a unit of thought. She remarked that:

> The documentary unit tends to draw close to the elementary idea of the unit of thought as forms of documents multiply, as the mass of documents grow and the professional technique of the documentalist is perfected.[17] (Briet, 1951, p. 10)

She therefore urged the documentalist not to be concerned only with the acquisition and analysis of documents but also with the communication of thought. Briet also stressed the necessity for the documentalist to produce value-added synthetic documents (analytical and critical summaries) in order to guide specialists and scholars in rapidly gaining an insight about new discoveries and results in their field of study by furnishing them with "analytical and sometimes critical summaries of new things of interest", if necessary by mastering computer technology (called "mecanography"[18] at the time) and statistical indicators.

Briet's documentalist is therefore an indispensable assistant to the researcher, knowledgeable in the subject matter the scientist or specialist is working on. As intellectual work that is not simply about producing abstracts of books or about finding them, documentation and documentalists can also arrive at entirely new creations "by juxtaposition, selection and comparison of documents and production of auxiliary documents. The content of documentation thus becomes interdocumentary". This statement published in her 1951 manifesto foreshadowed the emergence of scientific and technological watch and of bibliometrics. Briet stressed the need to acquire new skills and to analyse grey literature that had been neglected by traditional librarianship. This emphasis on producing timely critical analysis of publications emphasised the importance of journals for the rapid dissemination of research results and of scientific and technical information (STI).

Briet argued that the information profession could not turn its back on technology and should embrace it. This is all the more surprising given that digital tools were still in their infancy[19] at the time and must have been daunting to all but the engineers or scientists from the "hard" sciences (mathematics and physics). Briet's emphasis was that documentation should be a dynamic process focusing on information gathering and delivery, not on cataloguing and storing books.

---

[17]Our translation from Briet's original French text "*L'unité documentaire tend à se rapprocher de l'idée élémentaire, de l'unité de pensée, au fur et à mesure que les formes de documents se multiplient, que la masse documentaire s'accroît, et que la technique du métier de documentaliste se perfectionne*" (Briet, 1951, p. 10).

[18]This would have included (or been mainly) punch card equipment at the time.

[19]There was at the time, a growing use of punch card technology (sorters/collators/counters), notably in census record analyses from about 1900 onwards.

She also argued that documentation techniques should be taught to everyone such that pupils will be able to support their quest for knowledge and for truth with relevant information. As Maack (2004) observed, Briet can be considered a precursor of information literacy programmes which later became prevalent in many countries. Every secondary and high school in France has since been equipped with a documentation and information centre [*Centre de Documentation et d'Information* (CDI)].

Briet advocated greater proximity of professionals with their end users as well as a greater reactivity in the face of the evolution of technology. She was critical of the Universal Bibliographic Repertory [*Répertoire Bibliographique Universel* (RBU)] which was the major *oeuvre* of Otlet and La Fontaine's IIB. The RBU aimed to provide a bibliographic record of all the publications since the beginning of printing. It "contained two main files: An author file and a classified subject file, plus a range of subsidiary files necessary for managing the main files" (Rayward, 1997, p. 291). Between 1895 and 1930, the RBU had assembled about 16 million records of publications on cards. Briet asserted rather disdainfully that "documentology lost nothing by unburdening itself of the Universal Bibliographic Repertory that the whole universe has treated as a chimera and which does not offer a comparable interest to the most localised of collective catalogs".[20]

The Universal Decimal Classification (UDC) of which the first edition was published in 1905 provided a thematic classification system for the RBU. Each RBU card thus had a UDC number which enabled interoperability of library classification systems worldwide. As stated at the beginning of the book (cf. "Scope and aims"), the UDC was an enhanced version of the DC, created in 1876 by the American librarian Melvil Dewey and widely in use especially in the US and UK. Otlet and La Fontaine obtained permission from Dewey to develop his system. However, their recommendations not having been adopted by the DC staff resulted in a variant system, the UDC.

Briet advocated the building of more flexible and *bespoke* collections, classification schemes, abstracting and indexing services that put the end users' interest at their centre. Her vision was that of a service where users will be less dependent on librarians and can circulate freely amongst better organised library collections. This vision proved prophetic: the world has witnessed the emergence of documentary networks from the 1960s both at national and international levels, the invention of the Internet in the late 1960s, search engines from the 1980s and the first digital libraries in the 1990s. All these inventions have effectively disintermediated users from

[20]In the original French text, "*La* documentologie *n'a rien perdu à s'alléger d'un Répertoire Bibliographique Universel que l'univers entier a traité de chimère et qui n'offrait pas un intérêt comparable au plus localisé des catalogues collectifs*" (Briet, 1951, p. 9).

their dependencies on information professionals (librarians and documentalists), perhaps not the result Briet wanted.

Briet also worked towards fostering cooperation between library and documentation associations. Maack (2004) cited her 1953 article in which Briet defended a broad and unified view of information professionals. For Briet, librarianship, documentation, museography and archival administration were the four branches of the same field and body of knowledge. The four relied on the same techniques of cataloguing, indexing, classifying and were motivated by the same goal of delivering documents and information to people. The *differentia* lay in the type of document that each specialty predominantly deals with and in their institutional framework. Briet held leadership positions in international and national organisations. In addition to having co-founded the UFOD, she was vice president of the FID and was active in the IFLA and in the UNESCO. She headed the committee of the International Standardisation Organisation (ISO) on Documentation and contributed to developing the French cataloguing codes based on her work at the BNF (Buckland, 2005). In 1950, her contribution to the emergence of documentation was recognised by a cross of the "Legion of honour" [*Légion d'honneur*] which was presented to her by Julien Cain, her BNF director.

Although Briet did not use the terms "information" and "information science" in her writings, the problems she was addressing were those of Information Science. For Briet, "documentation" and "documentalist" (i.e., "he that documents the other"[21]) were umbrella terms encompassing the information profession and professionals (Maack, 2004).

1.3.7. Suzanne Briet: Posthumous Recognition Outside France

The proverb "none is a prophet in his country"[22] sums up the posthumous treatment reserved to Briet in her native country. A few publications recognised her contributions to the emergence of documentation in her time, especially in the journal *La Revue du Livre* edited by Georgette de Grolier and in the *Revue des Bibliothèques* (renamed *Archives et Bibliothèques*) edited by Henri Lemaître. Some homage has been paid to her by a handful of first-generation French scholars and practitioners in LIS (Sylvie Fayet-Scribe, 2000). However, from the 1980s, Briet's contributions to IS in France largely

---

[21]In her original words "*celui qui documente autrui*".

[22]This saying is taken from a Bible passage in the book of Matthew 13:53–57 entitled "A prophet without honour" where Jesus was preaching in his home town Nazareth and people who knew his humble beginnings were at first amazed, then angry and envious of his miraculous powers. Jesus replied to them: "A prophet is not without honour except in his own town and in his own home" (verse 57).

fell into the ashes of history. The generation of IS scholars who set up the first university courses in documentation completely ignored her contributions to the field. Her writings and works are noticeably absent from the higher education curricula in France such that it is possible to complete a whole cycle of university degrees in information-documentation without ever encountering Suzanne Briet's name.[23] This is largely due to a disconnect between the pioneers of documentation in Briet's time who were mainly professionals, and the founders of the discipline called "Information and Communication Sciences" who were academics from the humanities. The peculiar context in which IS emerged in the French higher education accounts for this "black hole" on the historical foundations of LIS (see Section 1.4.2 hereafter).

Fortunately, Americans can be relied upon to rediscover and celebrate neglected European masters. Briet was not the only one to have suffered such neglect or rejection in her native country. Paul Otlet's immense influence on the development of LIS was brought to light thanks to his dedicated biographer, the Australian scholar Boyd Rayward. The emeritus Professor of IS in France, Jacques Perriault (2007) also observed that the works of Abraham Moles (an expert on Claude Shannon) and of Pierre Schaeffer were better received abroad than in France. André Vitalis (2007) equally noted that the works of the French sociologist Jacques Ellul was for a long time ignored and rejected in France whereas he was "celebrated" in the US. "The Ellul Forum" is edited by the "International Jacques Ellul Society" based in the US. In the area of arts and culture, a more disgraceful fate befell the great French master of impressionism Paul Cézanne who during his lifetime was shunned from his home town, Aix en Provence. It took the dedicated efforts of some American entrepreneurs and philanthropists to restore Cézanne to his full glory and "return him" posthumously to the Provence.[24]

---

[23]The author of the current book only heard about Suzanne Briet during a fortuitous meeting with Emeritus Professor Michael Buckland of the University of Berkeley at an ISKO international meeting held in Montréal in 2008. This was 11 years after she had completed her PhD in Information Science in France! Laurent Martinet who co-authored the translation of Briet's manifesto into English was formerly a student at the INTD, thus in the very documentation school that Briet created. He had also never heard of Briet during his training in that institution. It was also a fortuitous meeting with two American LIS scholars – Michael Buckland (again!) and Ronald Day during a conference at his *alma mater*, the CNAM, that brought Briet to Martinet's attention. His testimony can be found in the preface to the electronic version of her book available at http://martinetl.free.fr/suzannebriet/questcequeladocumentation/.

[24]An exhibition by the Musée Granet in Aix en Provence in 2017–2018 bore the evocative title "Cézanne at home" and recalled this embarrassing story of the rejection of the "son of land" by the curator of the Granet Museum at the time who was quoted as having sworn that "Me alive, no Cézanne will enter the museum" ("*moi vivant aucun Cézanne n'entrera au musée*"). For more details, visit http://www.museegranet-aixenprovence.fr/expositions/actuellement/cezanne-at-home.html#.

To return to Briet, since 1995, Boyd Rayward (the rediscoverer of Paul Otlet) and Michael Buckland have led a worldwide and largely successful effort to bring Briet's contribution to the attention of the LIS community. Buckland's website hosts a compendium of articles and scientific events celebrating the contribution of *Madame Documentation* to IS.[25] The premier learned society in LIS, Association of Information Science and Technology (ASIST), the scientific editor of *Journal of the American Society for Information Science & Technology* (JASIST), published a homage to Briet, authored by Ronald E. Day (2007), Professor of LIS at the University of Indiana. The Special Interest Group (SIG) of the ASIST on the History and Foundations of Information Science (HFIS) has adopted the antelope as its emblem in homage to Briet's "antelope-as-document" example. Mary Niles Maack, Professor at the University of Los Angeles in California (UCLA) wrote an excellent biography of Suzanne Briet (Maack, 2004).

Briet's manifesto, *Qu'est-ce que la documentation?* has been translated into English and Spanish. If her manifesto captured the interest of LIS scholars on the other side of the Atlantic, it is because they saw in her analysis a visionary announcement of the transformation of the discipline and of its oldest institutions – libraries – towards special documentation centres (or special libraries in the US). Of particular interest is the broad perception that Briet had of documentation as a new specialty, endowed with social and cultural dimensions. Briet had stressed the necessity for the information profession to be proactive, in the face of "rhythms" of changes in technology. In this sense, her vision questioned the hegemonic role of libraries and archivists as sole custodians of knowledge.

Briet perceived documentation as a "network of production" and of "cultures of expression" thus anticipating what will be later known as the actor network theory. She also predicted the end of the "culture of the book" of which a result will be the transformation of traditional libraries into documentation centres.

In the preface to the English translation of her book, Ronald Day explained the singularity of Briet's view of the documentary function in 1951:

> What is documentation? is a revolutionary book in the field of library studies. It goes beyond Otlet's emphasis on books as the trope and cornerstone for documentation. It offers a vision beyond that of libraries and books, sees in documentation an unlimited horizon of techniques and technologies (and of "documentary agencies" employing these) in the service of the multitudes of particular cultures.

[25]Available at http://people.ischool.berkeley.edu/~buckland/briet.html.

In these regards, Briet's Manifesto remains a necessity of our time. (Day, 2007, Preface, v)

Buckland (2018) further highlighted the theoretical implications of Briet's radical ideas of what a document can be, that is, not only texts or graphics but concrete objects (stones and animals) used as learning objects in museums, for instance. He observed that this meant that Briet was working from a semiotic or phenomenological viewpoint rather than a symbolic or syntactic one which was usually the case in document analysis where only texts and written artefacts were considered "documents". Indeed, Briet's document can be seen to encompass four dimensions simultaneously:

(1) ontological: a document is materialised on some form of medium and is made up of signs and symbols;
(2) teleological: a document is created to serve as evidence in support of something, therefore intentionality precedes its ontological existence;
(3) semiotic: the signs and symbols inscribed on a document can be interpreted according to a semiotic process by a receiving mind; and
(4) phenomenological: a document must be perceived to be a document by another person and ultimately its interpretation lies with the perceiver.

This put Briet way ahead of her contemporaries:

Jean Meyriat and Jesse Shera probably thought that they were being progressive when they expanded the notion of document (and the scope of bibliography) beyond printed material to include other forms of graphic material, by which they meant physical expressions of human ideas. But it was a limited change. Briet's examples, in contrast, are a radical departure. They are simply not graphic, although one could debate the status of the photograph of a star. Unlike Meyriat and Shera, Briet and Pagès argued from a theoretical position, a semiotic or phenomenological position, a point that Shera simply did not comprehend. Jean Claude Gardin continued Pagès' ideas, but Meyriat did not. (Buckland, 2018).

Buckland (2018) went on to "suggest that the (much-delayed) working-out of the implications of Briet's position – a neo-documentalist agenda – is one of the most important conceptual developments at this time in library and information studies" since "Significant intellectual developments in the information sciences are very rare."

Another more fundamental implication of Briet's radical ideas about the nature of documents is that it places emphasis on context when analysing documents. The context of the production of documents has been neglected in the LIS field where assumption has been that a text has only one meaning regardless of the context in which it is placed and regardless of the audience contemplating it. The implication is that LIS has implicitly based its methods for indexing and classification on the assumption of the immutability of signification of written artefacts:

> The entire field of bibliography, documentation, knowledge organization, and the sciences of communication and information have been dominated by notions of sameness and transmission. Mass communication is concerned with the delivery of the same message to many people. The preservation of recorded knowledge through time and space is central to all varieties of document theory. Bibliography is very much concerned with determining whether two items are the same or not. Shannon's information theory and textual studies are both concerned with assessing and reducing "noise" and corruption. (Buckland, 2018).

Buckland's point is that once an object has been detached from its original context and placed in a different context, this influences the comprehension and meaning of the said document. He went on to observe that:

> "Existing theory in knowledge organization and information science is dominated by assumptions of sameness, especially in the role of documents in mitigating the effects of time and space. Although Briet does not appear to address this point, her examples suggest the opposite. All three examples are of irreversible detachment, of physical change associated with temporal and spatial separation from their original context. Only in the rarefied extremes of mathematics, logic, and computational theory are time and place erased and processes considered reversible.
>
> There is an important corollary: because documents are material objects (whether paper, microfilm, or electronic), if they are removed from a context, they must inevitably arrive in another new and different context. As Derrida said, "*il n'y a pas de hors-texte*" [There is no being out of context] [DER 67, 227]. Neither we nor documents can be contextless. This is important because context influences meaning and interpretation. But if this aspect has been largely overlooked in our field, it has been noticed outside our field. [...] The implication is that our whole field has been looking in the wrong direction, or, at least, casting a gaze that is importantly incomplete. There is research on the information seeker and the seeker's context. It seems we need a complementary research agenda on the document and its context." (Buckland, 2018)

If we understand the acuity of Buckland's observation about the importance of considering the different contexts in which a document is set, it seems equally important to put a framework in place to guard against an extreme form of contextualisation which could lead to an excess of constructivism and relativity resulting in a form of "anything goes" and "any interpretation can be valid". The theoretical and epistemological assumptions underlying such context-bound interpretations need to be spelt out beforehand.

With worldwide interest re-ignited on Briet's writings, thanks to her American admirers, the field of IS is witnessing a *renaissance* of the documentation movement. However, it is fair to point out that although Briet's ideas appear advanced for her time, she was not the only one to uphold documents and documentation at the time. The seeds of her ideas were already sown in Otlet's 1934 *Treatise on documentation* and elsewhere. Nevertheless, Briet and her antelope have captured people's imagination worldwide like no other scholarly discourse on document has. Generations of LIS students have been assigned readings about her.

The serendipitous meeting of three scholars Michael Buckland, Boyd Rayward and Niels Lund Windfield at the second COLIS conference in Copenhagen in 1996 resulted in the founding of the Document Academy,[26] the scholarly movement which upholds the centrality of documents over information and bibliography. A more transient "document *renaissance*" took place in France around a multidisciplinary group of researchers from Computer Science and the Humanities which was formed in 2003 with funding from the French National research Center [*Centre National de Recherche Scientifique* (CNRS)] to investigate the nature and implications of the digitisation of documents. Under the collective pseudonym of Roger T. Pédauque, this group published a manifesto book in 2006 on the transformations of the document in the digital era. The group has now been disbanded.[27]

A noteworthy attempt to honour francophone pioneers of information and documentation was made during the 11th colloquium of the French chapter of the ISKO held in Paris in 2017. Several papers were devoted to French pioneers, including Suzanne Briet, Eugène Morel, Jean-Claude Gardin and Robert Estivals.

These belated attempts to acknowledge francophone pioneers of IS in France is all the more ironic given the considerable influence they had in shaping LIS in other European countries, especially in Latin Europe (Spain, Portugal), in Latin America and in the US. Aside from Paul Otlet and Suzanne Briet, other French LIS pioneers were well known by their European contemporaries. Jean Claude Gardin was well known internationally and published quite a bit in English. In 1962, he was invited by Björn Tell to give a talk in his new course on informatology at the Swedish Royal Institute of Technology at the time when Gardin was developing his

---

[26]Website http://documentacademy.org/.

[27]The collective book written by members of this group is entitled *Le document à la lumière du numérique, C&F éditions, 2006*. The story of the Pédauque group can be found here: http://base.d-p-h.info/fr/fiches/dph/fiche-dph-8220.html.

SYNTOL system. Eric de Grolier and Eugène Morel were also well known by their European contemporaries.

1.4. From Documentation to Information Science

While Paul Otlet, Henri La Fontaine and Suzanne Briet undoubtedly laid the conceptual and organisational foundations in the pre-mechanised era, the road to the official recognition of IS in the French higher education system was not straightforward. There was no direct line going from Otlet, La Fontaine and Briet to the academic discipline IS in France. The post-Briet period was marked by the development of research in information processing worldwide. This signalled a shift in focus, from the form or container (documents) to the contents (indexing and retrieval). The study of the former was relegated to *bibliothéconomie* (library economics) and to librarians, while the latter became the research theme of a new batch of scholars with a more formal background. This section borrows from the author's earlier publications on the state of IS in the French higher education landscape (Ibekwe-SanJuan, 2012a, 2012b, 2012c, 2012d).

1.4.1. The Beginning of Mechanised Documentation

Some of the prominent figures in this transition period were Robert Pagès, Jean-Claude Gardin and Eric de Grolier.

Buckland (2017b) remarked that Robert Pagès (1919–2007) had had many lives: he started his adult life as a left-wing political activist – he was a Trotskyist and a clandestine anarchist activist. He studied psychology in Paris in 1945 and in 1946 enrolled as a student in Briet's documentation programme at the UFOD. His interest in documentation problems permeated his later scientific career at the French National Center for Scientific Research (CNRS) which he entered in 1951 and where he founded a major social psychology research laboratory [*Laboratoire de Psychologie Sociale*].

Pagès was critical of the rigidity of universal classification of languages (notably of Otlet's UDC and Melvil Dewey's DC) which forced a document to be classified under only one subject, thus into a monothematic class. He considered that the focus should be on analysing and retrieving content, not organising books. He therefore advocated an *n*-ary dimensional analysis of the contents of documents and tried to formalise the relation between documents, books and experiences.

His idea was that a document is an instrument for accumulating symbolic activity; hence, a document is made of signs and symbols that are subject to interpretation. These symbols acquire meaning outside of their context of production. He explored other types of symbols like mathematical language and advocated the grounding of documentary languages on a formal scientific basis. This led him to create a formulaic representation of the content of documents which he called "coded analysis" [*analyse codée* (CODOC)]. CODOC went into operation in 1954 and was deployed in the library of his psychology research laboratory. The idea behind CODOC was to design an extensible grammar and a lexicon that would enable the creation of new and unexpected classes, thus giving an infinite possibility for subdivision while being easy to memorise (i.e., having a mnemotechnical faculty). However, the result was an artificial language for indexing and classification that was inhospitable to memorisation (Ibekwe-SanJuan, 2012b).

Robert Pagès's defence of a "science of documentation" which he called documentology [*documentologie*] was as visionary as Briet's semiotic conception of document was in 1950. For Pagès, documentation was a larger specialty subsuming library management because the latter was about books whereas the former was about documents, a category that subsumed books. Much in the same way that Briet thought documentalists should be the assistants of scientists, Pagès thought that "documentology" should be part of symbolic communication and should be populated with "researchers-cum-documentalists". This would ensure that documentation is not reduced to a set of techniques aimed at solving practical problems. The term "documentology" was unsuccessfully taken up later by Robert Escarpit, one of the founders of the Information and Communications Interdiscipline (ICS) and also by Jean Meyriat (1981a) in his attempts to name the field of IS within the ICS interdiscipline (see Chapter 5, Section 5.2.3 for a discussion).

Decidedly, Briet and Pagès's ideas are remarkably too close for it to be pure coincidence (see Buckland, 2018). Both held a wide semiotic conception of documents as being anything including humans in Pagès's case, insofar as it helps to prove the existence of a physical or abstract entity. As mentioned above (in Section 1.3.5), Pagès's conception of a document was first published in 1947 in his *mémoire* "*Transformations documentaires et milieu culturel (Essai de documentologie)*" and reproduced as an article in 1948, thus three years before Briet's *Qu'est-ce que la documentation?* It would then seem that Robert Pagès should share the podium with Suzanne Briet as one the founders of French documentation and a precursor of IS. Buckland's recent publication on the connections between Briet and Pagès has already sparked a renewed interest in Pagès amongst IS scholars in France.[28]

[28]Olivier Le Deuff an IS Professor in Bordeaux recently began revisiting Pagès's writings. He wrote an online article in April 2018 on Pagès's contribution to documentation which prefigured the emergence of IS. Retrieved from https://www.reseau-canope.fr/savoirscdi/societe-de-linformation/le-monde-du-livre-et-des-medias/les-penseurs-de-linformation-de-la-documentation-et-de-la-pedagogie/robert-pages-une-necessaire-redecouverte-par-les-sciences-de-linformation.html. Accessed on 24/04/2018.

Another prominent figure of this transition period, Jean-Claude Gardin (1925–2013) also had his career at the CNRS. Gardin was an archaeologist confronted with the problem of sorting and comparing archaeological objects referred to in scientific texts. Wishing to reduce the labour-intensive nature of retrospective research and to systematise the conceptual analysis of the contents of scientific communication, he designed a sophisticated system for facet analysis, indexing and automatic document retrieval called SYNTOL (Syntagmatic Organization Language) in 1964.

CODOC and SYNTOL shared several common features: they were designed by two CNRS researchers from the Social Sciences; they adopted a formal language for indexing which was new at the time and marked the transition from bibliographic analysis (classification of books) to content analysis (indexing and information retrieval).

Gardin was known for his research on discourse analysis applied to archaeology but he was critical of the emerging fields of natural language processing (NLP) and artificial intelligence (AI) and especially of their promises of automated semantic analysis of natural language. He believed that such systems can work only in very restricted domains with a well-circumscribed terminology. History has proven him right. Attempts at building large coverage semantic parsers have failed and have been overtaken by search engines that rely on surface pattern-matching, probabilistic models and the large-scale leverage of the link structure of the web (the PageRank algorithm exemplifies the success of such algorithms).

It is noteworthy that the theoretical discourse and aspirations of Gardin and Pagès were somewhat at odds with their attempts to create a formal language for content analysis. The formalisation of semantic relations ultimately leads to reducing the rich nuances which make up human language and therefore to a narrow and physical conception of human language (Ibekwe-SanJuan, 2012b).

Also, the level of sophistication and human expertise required to implement CODOC and SYNTOL made them intractable to implement on a large scale. Their capacity for handling complex relationships made them complicated to use (both for the indexer and the searcher) and the triviality and cost effectiveness of a simple boolean string search approach rendered these first-generation syntactic and semantic indexing systems obsolete. Hence, while Gardin and Pagès laid the foundations in France for the development of research on automatic indexing and information retrieval, it was ironically the arrival of computers and search engines that blew their systems and other powerful indexing systems out of the water.

Other important figures deserving mention in this transition period are the de Groliers (1911–1998) husband and wife. Eric de Grolier, Georgette de Grolier and Suzanne Briet worked towards the development of documentation and for a *rapprochement* between the two rival professions of librarians and documentalists. Eric de Grolier took part in efforts to set up the first training courses for documentalists at the UFOD in the 1930s. He sought for ways to make specialised classification languages more compatible through normalisation of classification schemes. He was thus preoccupied with the interoperability of classification languages. Since the advent of the Internet and the web, the interoperability of classification languages has become a concern for many scientific communities including IS. Research and initiatives on this theme have led to the development of norms and standards to describe electronic documents such as the Text Encoding Initiative (TEI), Dublin Core and SKOS (Simple Knowledge Organization System). These norms for metadata encoding have given documentary languages a *renaissance*, enabling them to continue to claim some relevance in the web era where all the information-seeking activities have become reliant on digital platforms.

1.4.2. The Rivalry between Professional LIS Associations in France

Fayet-Scribe (1997) studied the roles played by three professional associations in the interwar period (between First World War and Second World War) in organising the library and documentation sectors. She researched the roles of the *Bureau Bibliographique de France*,[29] the French arm of IIB founded in 1902 by Général Hippolyte Sebert (see Section 1.3.1 supra), the UFOD and the "Association for the development of public libraries" [*Association pour le Développement de la Lecture Publique* (ADLP)].

The UFOD had been co-founded in 1931 by Suzanne Briet and Jean Gérard with the support of Julien Cain. Its main objective was to foster cooperation in documentation and information management in corporate organisations. Membership was open to both private and public organisations and to individuals from all sectors of activities, including ministries, scientific and industrial organisations. In 1932 and under Briet's aegis, the UFOD undertook the task of defining a documentation vocabulary aimed at providing documentalists with a common language. The desire for such a common language had been expressed during the 1910 congress of the IIB. In 1933, Briet presented an embryonic lexicon translated into English and German which helped to establish an identity for documentation now seen as a separate field from bibliography and librarianship (Delmas, 1992, p. 186).

[29]Not to be confused with the *Bulletin des Bibliothèques de France* (BBF), the leading professional journal for librarians in France.

The mission of the ADLP was to create a network of public libraries that were available to all.

There was a significant overlap in the membership of the three associations. The *Bureau* included members of the Association of French Librarians [*Association des Bibliothecaires Français* (ABF)] such as "Eugene Morel, Henri Lemaitre (1881–1946), Emest Coyecque (1864–1954), and Gabriel Henriot (1880–1965)" who were also teachers at the *Ecole de Chartes*. Briet and Henri Lemaître had key roles in the UFOD but were also active in the BBF and the ADLP. The De Groliers (Eric and Georgette) were at the intersection of the three associations as the *Bureau*'s secretaries, UFOD affiliates and founders of the ADLP in 1936.

As well as being close collaborators of Suzanne Briet, the De Groliers formed "one of the most active reformist couples in the library/information domain during the interwar period" Fayet-Scribe (1997, p. 786). The *Bureau* and the UFOD worked closely for several years to foster awareness of new methods of information handling, especially of scientific and technical information (STI) and of documentation.

Delmas (1992, p. 188) listed several reasons for this good "*entente*" between the professional associations of librarians and documentalists in the period before and after Second World War. Firstly, the two professions were not yet fully structured, the contours of each were still imprecise and documentation was seen as a melting pot where all those who wished to modernise the very conservative outlook on librarianship could experiment with new methods and techniques. Secondly, many of the important figures (Eugène Morel, Armand Boutillier du Retail, Henri Lemaître, Charles Sustrac, Suzanne Briet, Louise Noëlle Malclès, Eric and Georgette de Grolier, etc.) were quite versatile. They were not only advocates of the new documentation techniques, they were also experienced librarians interested in public literacy issues. Thirdly and perhaps more importantly, they knew and respected one another and were united in their efforts to cultivate the emerging field of documentation. Finally, information technology (computers) was still in its infancy at the time and contributed equally to the two professions.

However, as the roles of the *Bureau* and the UFOD became more and more similar, rivalry set in. Finally the *Bureau* was dissolved in 1950.

Hence, after Second World War, an atmosphere of rivalry and hostility permeated the information professionals association:

> "The most significant contacts between documentalists and librarians in France took place during the 1895–1944 period. After World War II, however, there was a distinct breach between the two." (Fayet-Scribe, 1997, p. 791)

Palermiti and Polity (2002) also observed that before World War II (WWII), "Documentation centres and libraries co-existed, the latter sometimes being at the origin of the former and housing it within its walls. Pioneering figures such as Suzanne Briet, Louise-Noëlle Malclès, Yvonne Oddon, Myriem Foncin et Georgette De Grolier were versatile, having worked in either structures and knowing them well. The two professions have the same history even if debates on documentation and libraries or precisely on the nature of documents arise now and then."

The creation in 1963 of the French Association of Documentalists and Special Librarians [*Association Française des Documentalistes et Bibliothécaires Spécialises* (ADBS)] was a reaction to the creation of the ENSB (*Ecole nationale supérieure des bibliothécaires*) in 1963, seen by the documentalists at the UFOD and the INTD[30] as a rival institution (Delmas, 1993, p. 222).

Such rivalry between professional associations is not peculiar to France nor to the LIS sector. Williams (1997) pointed out the tense relations between the ALA and the American Special Libraries Association (SLA). The creation of the SLA in 1909 was the result of a split from the ALA which was founded in 1876. The word "special" in the name of SLA raised much animosity from the ALA and general librarians who retorted that "all libraries are special". The SLA retained its name because "documentation" was not appealing and it could not come up with a more appealing name (Williams, 1997, p. 776).

Maack (2004) observed that owing to the weight of history, library modernists in France could not call their new documentation centre by the term "*bibliothèque spécialisée*" which would have been the French translation of "special libraries". A new term had to be used. The SLA had a clear filiation with the European documentation tradition laid by Otlet, Briet and the FID. Citing Aksel G. S. Josephson, Williams (1997, p. 776) wrote:

> "We might come to the point where special libraries will not even have whole books, but only such parts of many books as it needs, treating books as well as periodicals on the principle of documentation (SLA Meeting Summary, 1912, pp. 147–148)", (Aksel G. S. Josephson). Their user-oriented services made the American special librarians the "first American documentalists".

Watson Davis, the founder of the American Documentation Institute (ADI) in 1937 was also influenced by the European tradition of

[30]INTD's *alumni*, in French, the "*Association des anciens élèves de l'INTD*" (AINTD) were active in the creation and in the membership of the ADBS.

documentation which was concerned not only with books but with all sorts of documents and served its users in places other than libraries, using the "cutting edge" technology of the time, that is, microfilms (Williams, 1997, p. 777).

Initially, the SLA and ADI had close ties. There was also a significant overlap in their memberships, up to 67% in 1961, according to Williams (1997) but gradually, there was a feeling on the part of SLA that the newly created ADI and the documentation movement were superfluous.

More importantly, SLA suffered from many ills: its reluctance to embrace the new information technology and the lack of specialised training programme for special librarians most of whom came from the business and industry sectors. The result was that it fell into decline as ADI rose to prominence and later became the Association of the American Society for Information Science (ASIS), then later ASIST (with T for Technology).

The Netherlands also had a special libraries association created in 1921 (Delmas, 1992, p. 186). In the UK, documentation centres were being called "information bureaux" and were organised by the Association of Special Libraries and Information Bureaux (ASLIB) founded in 1924.

1.4.3. The Emergence of Information Science and the Disconnect with Professional Associations in France

In contrast to what happened in the UK and the US, IS emerged in the French higher education without input from two of its major professional bodies – librarians and archivists. Only the documentation arm was somewhat associated with the creation of IS in France. Calenge (2000) offered a summary of the three major fault lines (*ruptures*) that opposed librarians and documentalists:

1. A major bone of contention is the difference in their educational institutions. Librarians and archivists are trained in special schools outside of universities which are run by professionals librarians and archivists. These special schools enjoy State recognition and have a homogeneous standardised curriculum. However, some librarians trained by local governments have a less regulated and less rigourous training and therefore a less secure career prospect. Documentalists and other information professionals on the other hand, are trained via professional associations of documentalists and mainly in technical colleges [*Instituts Universitaires de Technologie* (IUT)] affiliated with universities. IUTs offer one to two-year technical degree programmes. In general, the training received by librarians, whether through the national programmes or at the local level tends to place more emphasis on

collection management and on public literacy [*lecture publique*] while the educational programmes of documentalists and other information professionals tend to be more technologically oriented and diverse. The fundamental issue raised by these demarcations in training institutions for librarians and archivists on the one hand and documentalists and other IS professionals on the other is the reluctance by the French government to let the state-funded universities handle the training of its elite (Palermiti & Polity, 2002). The distinction between universities that are obliged to accept practically all freshmen candidates and the *Grandes Ecoles* that train the country's elites is one of several *exceptions françaises*. The situation of the ENSSIB is quite peculiar in that it offers both the nationally accredited training programmes for librarians and also several university degree programmes for documentalists and information professionals trained at the same high level as the Parisian INTD. Thus, the ENSSIB is one of the few hybrid LIS institutions in France that houses three LIS educational programmes under one institution: documentation, librarianship and IS, although as Calenge (2000, p. 19) underlined, the three programmes are separate and run parallel to one another.

2. Once they have obtained their degree, librarians and archivists become civil servants in the national public service [*Fonction publique d'Etat*] which guarantees them job placement. Librarians trained in the national library schools work in national, university libraries and in public information libraries while those trained at the local level can only work in departmental and municipal libraries managed by towns and local governments. Their career is therefore wholly dependent on the local government and less open to mobility. Documentalists face an even more precarious career in the private sector.
3. These grievances are carried over into their professional lives where each information professional has his or her own association, and publication venues. The French National Library Association [*Association des Bibliothécaires de France* (ABF)] publishes the *Bulletin Bibliographique de France* (BBF) founded in 1956 by Julien Cain, then Director of the BNF and mentor of Suzanne Briet. Documentalists and the other information professionals congregate around the ADBS which founded the journal *Documentaliste, Sciences de l'information* in 1963 the same year of its creation. This journal began as mostly a professional publication. However, a reshuffle in its editorial policy and board led to the integration of some academic content and of a few IS scholars. Since 2015, this journal has been renamed *I2D – Information, Données & Documents* (Information, Data and Documents) and the ADBS has been renamed *Association des professionnels de l'information et de la documentation*. The incorporation

of the words "information, data and document" reflects a desire to align itself with the core concepts of IS.

The orientation of the education programmes for librarians in France has been subject to recurrent debates. Opinions are divided between two options: either train "generalist" librarians who master documentary techniques or train subject specialists in library management. According to Calenge (2000, p. 23), the former has been the case because "the French tradition erected library techniques as an end in itself, which in university libraries led to a relative lack of understanding between IS scholars who are preoccupied with knowledge transmission and the librarians that uphold their techniques as knowledge in itself."

Another reason for the disconnect between LIS professionals and the academic field in France is the fact that librarians and archivists do not require university degrees to work nor are they expected to carry out research. They have existed as separate professional bodies concerned only with managing libraries and archives and not preoccupied by research on these institutions. There is no incentive for them to be involved in university education programmes as this is not valued in their career progression (Calenge, 2000, pp. 19–20). The consequence was that "the academic recognition of the discipline was done on the borderline of its already established cognitive institutionalisation: only a small proportion of its heritage was associated with the academic constitution of IS, leaving by the roadside a part of the public institutions (libraries, museums, archives) and some researchers who had invested time on IS research" (Palermiti & Polity, 2002[31]).

1.4.4. Impact of Governmental Policies on the Emergence of Information Science

The current state of IS in the French higher education landscape is a result of an incoherent, technocratic and instrumental policy on Scientific and Technical Information (STI) implemented by the successive French governments between the early 1970s and the late 1990s. Salaün (1991, 1993) documented how LIS field was reduced to policies on STI and instrumentalised for national strategic and geopolitical goals. This affected negatively the theoretical development of the field and hence its ability to legitimise itself as an academic field.

---

[31]Our translation of the original article in French.

Much in the same way as the American government reacted after the launching of the Sputnik satellite by the USSR in 1957, the Gaullist[32] party which governed France after Second World War saw STI as a tool with which to advance its political agenda and gain independence from the two superpowers – the Soviet Union and the US.

Governmental policies on STI did not seek input from humanities scholars nor from professional associations of libraries and documentalists but rather from technocrats and scientists from nuclear physics, chemistry and engineering. This situation is not peculiar to France nor is it specific to the post-Second World War era. Indeed, scientists and especially chemists rise to prominency in times of war. Fayet-Scribe (1997, p. 784) observed that many scientists were closely associated with the Nazi's rise to power in Germany:

> After the end of World War I, in fact, there was an awareness in France of the necessity for a national effort in the realm of what we now call technical and scientific information. Thanks to its wealth of bibliographies, German science had created a monopoly situation and closure of the borders had halted the flow of publications, a penury that brought home the importance of access to information. (Gablot, 1991)

Successive French governments relied on different governmental agencies to implement their policies on STI. The French National Research Agency (CNRS), the Commissary for Atomic Energy [*Le Commissaire à l'énergie atomique* (CEA)] and the Defense ministry all played important roles. Each agency tendered a competing conception of what the nation required in terms of STI infrastructure. The CNRS advocated a centralised structure while other groups of experts advocated a decentralised one. The implementation of each policy was overturned at each change of power at the very top. The result was that the development of LIS was delivered to different political factions, each seeking to leave its mark on STI infrastructure and reverse the decisions taken by their predecessors from the opposing political party.

First, the Gaullist party that governed France for 15 years after Second World War devised a "grand plan for computation" [*plan calcul*]. This plan was based on the Boutry Report of 1964 aimed at addressing the lack of adequate storage and retrieval facilities to manage the exponential growth in document production. The *plan calcul* signalled a shift of focus from documentation to STI, technology and industry, thus preventing any real scientific debate on IS because STI was presented as an issue of national sovereignty (Salaün, 1991). This "grand plan" placed the documentation

[32]Political ideas assimilated with Charles de Gaulle's Republican Party.

centre of the CNRS in the role of coordinator of the national effort on STI. The CNRS created the "Analytic bulletin" [*Bulletin analtyique*] (renamed *Bulletin signalétique* in 1956) which provided structured bibliographic records of STI organised under three major scientific sections ("hard" and engineering sciences, life sciences, philosophy and humanities). Various ministerial and industrial sectors created or enhanced their STI documentation centres. In 1968, the "Documentation and STI service of the army [*Service de documentation scientifique et technique de l'armée* (SEDOCAR)]" was placed under the Department of Research and of military resources of the Army and renamed "The Documentation Center of the Army" [*Centre de Documentation de l'Armée* or CEDOCAR[33]]. Huge efforts were deployed to create bibliographic databases in physics and chemistry in order to gain independence from the monopoly held by the American Chemical abstracts and Inspec databases.

The arrival of the centrist party and the election of Valéry Giscard D'Estaing as president in 1974 brought a change of policy in STI. The *plan calcul* was abandoned in favour of the "telephone plan" [*plan telephone*], a vast project for the development of telecommunications infrastructures throughout rural France. This new project was launched by the landmark Nora-Minc report of 1978 (named after its authors) entitled "The Automation of society" [*L'informatisation de la société*]. This report compared information to energy, thus to a precious natural resource which the society was in dire need of. It was in this report that the neologism "*télématique*" was coined, a contraction of "infor*matique*" (informatics) and "*télé*communications". As Salaün (1991, p. 15) observed: "With this report, information gained a higher status and acquired its "lettres de noblesse", although its exact nature and role were never clearly defined." The telecommunication industry that was developing infrastructures in rural France became a prominent actor in the information sector and entertained a confusion between informatics, telecommunications and information. Information was not seen as something which had content, meaning and could be analysed but rather as something which can be transported, whose volume can be measured and whose influx can be observed. Databases, servers and big documentation centres were being compared to grey petrol (*pétrole gris*). This processual conception of information is in line with the Shannon-Weaver theory of mathematical communication where information is seen as a physical entity to be processed, transported and delivered from one point to another. Assimilated with informatics and "*telematics*", information was being presented as a means to bring France out of the

[33]According to a short summary given on the website of the BNF. Retrieved from http://data.bnf.fr/11862763/centre_de_documentation_de_l_armement_france/.

economic depression into which it had sunk in the mid-1970s. Information therefore became an industry and the term "information society" (*société de l'information*) became a popular buzz word of the 1980s. This contributed in obscuring the links between the emerging field of IS and its historical roots in librarianship and documentation.

With the arrival of the socialist president François Mitterand in 1981, another reversal in STI policy was witnessed. The socialist government focused its efforts on the democratisation of information and on scientific communication. In 1982, the minister of research and industry of the time, Jean-Pierre Chevènment, convened a national conference with the scientific community where the emphasis was laid on scientific publication.

Several ministerial agencies were created to advise the government on its STI policies. The *Commissariat general au plan* led the effort for the development of STI infrastructures. In 1973, it created the National Bureau for Scientific and Technical Information [*Bureau National d'Information Scientifique et Technique* (BNIST)] which was the first in a long series of government agencies whose abbreviations ended with STI (Salaün, 1993). The BNIST was the French counterpart of the "Committee on Scientific and Technical Information (COSATI) in the US" (Fayet-Scribe, 1997). An attempt to bring together the different ministerial bodies in charge of STI resulted in the creation of the "Direction for Libraries, Museums & Scientific and Technical Information" [*Direction des Bibliothèques, des Musées et de l'Information Scientifique et Technique* (DBMIST)] in 1982. This was followed by the creation of Centres for Acquisition and Dissemination of Scientific and Technical Information [*Centres d'Acquisition et de Diffusion de l'Information Scientifique et Technique* (CADIST)], then by the creation of the "Regional Training Center for Scientific and Technical Information" [*Unité Régionale de Formation à l'Information Scientifique et Technique* (URFIST)]. The National University Server for Scientific and Technical Information [*Serveur Universitaire National pour l'Information Scientifique et Technique* (SUNIST)] was created in 1984, followed in 1988 by the National Institute for Scientific and Technical Information [*Institut National de l'Information Scientifique et Technique* (INIST)], the STI arm of the CNRS. The INIST hosts the two national scientific bibliographic databases, PASCAL and FRANCIS. As Salaün (1993) observed, "this inflation of governmental agencies on STI signaled the shifting of focus from "biblio-" (as in bibliothéconomie") – thus from books and documents to – STI, a change of object and the fact that the focus of analysis is henceforth on technocratic issues".

This instrumental approach to information in the first decades of the existence of the academic discipline (see Section 1.4.5 hereafter for an account of how IS emerged in France) unfortunately overshadowed the wider conceptual approach to documents defended by Briet and Otlet. It oriented IS education and research towards document management and towards developing solutions for the information industry (telecommunications, databases and computer systems):

> Whereas at the beginning of the century our actors clung to the prefix biblio- to underscore the fact that a pluridisciplinary reflection was being applied to document, the abuse of "-IST" is testimony to a more technocratic approach and a change of object. The result was not the building of a science but of structures concerned not with documents but with information.[34] (Salaün, 1993)

These inconsistencies in the successive government policies affected the epistemological and conceptual emancipation of the field negatively. The result, as Palermiti and Polity (2002) aptly put it, was that the management of STI displayed an unfortunate characteristic trait of the French administration: "a jacobine[35] centralisation, a policy of "pré-carré[36]" undertaken by the different government agencies and a general absence of coordination between the different government bodies".

1.4.5. The Birth of an Interdiscipline: Information and Communication Sciences

As discussed previously, the line from Suzanne Briet, Robert Pagès, De Grolier's and Jean-Claude Gardin's pioneering works in documentation in the 1950s to the official recognition of IS as an academic discipline in the French higher education landscape was not straightforward. It was interrupted by incoherent governmental policies focusing solely on STI and aimed at developing French autonomy in information technology. This situation brings to mind an observation made by Roland Barthes that "*to build itself, the 'myth of origins' needs to forget its history*[37]" (cited in Jeanneret, 2007, p. 110).

---

[34]In the original text: *Alors qu'au début du siècle nos acteurs s'accrochaient au préfixe biblio- pour marquer que l'on appliquait une réflexion pluridisciplinaire au document, l'abus des "-ist" témoigne à la fois d'une démarche plus technocratique et d'un changement d'objet. On ne construit pas une science, mais des structures. Celles-ci ne traitent pas des documents, mais de l'information* (Salaün, 1993).

[35]This refers to the "*concept of a centralized Republic, with power concentrated in the national government, at the expense of local or regional governments*" (http://en.wikipedia.org/wiki/Jacobin_(politics)).

[36]"Pré-carré" means one's perceived turf or area of monopoly which is usually defended against newcomers and rivals.

[37]Barthès's original text is "*pour se constituer, le mythe des origines a besoin d'oublier son histoire.*" (Barthes, 1957, p. 203).

Several publications in French (Boure, 2002; Meyriat, 1993; Tétu, 2002) have given accounts of how the "Information and Communication Sciences" (ICS henceforth) interdiscipline [*Sciences de l'information et de la Communication* (SIC)] emerged in the French higher education landscape. The artisans of this interdiscipline were three men with a strong literary background: Robert Escarpit and Jean Meyriat "played a primary role, assisted occasionally by Roland Barthes" (Tétu, 2002). In 1972, the trio created a pressure group called the Committee for Information and Communication Sciences presided over by Jean Meyriat. This committee became the French Society for Information and Communication Sciences [*Société Française des Sciences de l'Information et de la Communication* (SFSIC)], the learned society representing this interdiscipline. The existence of the Committee on ICS and of the SFSIC paved the way for the official recognition of the ICS interdiscipline in 1975 by the French Ministry of Higher Education.

Robert Escarpit (1918–2000) was a versatile scholar with a background in English and in sociology. He was also a writer and a journalist. He regularly wrote for the French daily *Le Monde* and for the UNESCO. His "conversion" to communication studies came from his observation that one cannot write without being preoccupied with how the message is received by the readers. Thus, he sought to shift the focus in literary studies to what he called "communicology",[38] that is, the relation between the author and his readers. In 1960, he founded a Centre for Sociology of Literary Facts which was later renamed "Laboratory for Information and Communication Sciences" [*Laboratoire des Sciences de L'information et de la Communication*] in 1978, thus making it one of the first research centres of the new interdiscipline. The first educational programmes in ICS were set up in Bordeaux university in 1968 where Escarpit did all his education and career (Laulan, 2007). He went on to become a Professor of ICS and President of the University of Bordeaux.

An internationally renowned scholar in structuralism, semiotics, literature and philosophy, Roland Barthes (1915–1980) was an even more versatile scholar. He earned degrees in literature, grammar and philology as well as in linguistics, sociology and semiology. He gave talks in several foreign institutions and in 1977 was elected to the *Collège de France* (equivalent of the Academy of Sciences in the US) which was the crowning of his scientific career.

Jean Meyriat (1921–2010) had a post-graduate degree in literature (*agrégé de lettres*) from one of France's elite school, the *Ecole Normale*

[38]The term was already in use and is attributed to a certain Wendell Johnson (1958). See the International Communicology Institute for more details. http://www.communicology.org/content/%E2%8C%98-definition-communicology. Accessed on 9 January 2018.

Supérieure d'Ulm. He then went on to work at the National Foundation of Political Science which was the ancestor of the current Political Science School in Paris, Sciences Pô Paris. There, he created the first documentation centre of this institution which he managed from 1950 to 1960. In 1962, he became Director of studies and Chair of Scientific Information at the "School of Higher Education in Social Sciences" [*Ecole des Hautes Etudes en Sciences Sociales* (EHESS)] also in Paris. He created the first post-graduate program in documentation and IS in 1969 at Science Pô. He then became head of the Sciences Pô's "Center for International Study and Research" [*Centre d'Etudes et de Recherches Internationales* (CERI)] between 1952 and 1976. He was president of the SFSIC between 1972 and 1984 and played a key role in trying to legitimise documentation and IS within the ICS. He made numerous attempts to define and delineate the frontiers of this subfield of the ICS (see Section 5.2.3 for a discussion of his terminological *oeuvre*). Meyriat was also internationally visible: he presided the European Council of Information Associations (ECIA); he was actively involved with professional associations in LIS, in particular with the ADBS and the FID.

After the official recognition of ICS as a scientific discipline, a small bulletin called "The Inforcom Letter" [*La Lettre d'Inforcom*] began to appear from 1978, the year of the first congress of the SFSIC.

Tétu (2002) observed that apart from Robert Escarpit, the majority of the literary scholars who were present at the creation of the committee for the ICS did not join the new interdiscipline, "rather they worked on communication problems but from the perspective of their own discipline (sociology, semiotics, linguistics)". Hence, the creation of the ICS field was not the result of a consensus on its subject matter (ontological question), theories and paradigms (epistemological question) but was rather the result of an opportunistic coming together of scholars who were interested in communication but from the perspective of their own fields. This ontological and epistemological *faux pas* continues to dog ICS until today. It has largely failed to reconcile the multidisciplinary communities that inhabit its multiverse. As an interdiscipline encompassing several scientific communities (media studies, communication studies, cultural studies, LIS, etc.), each with its own tradition of research, theories and methods, ICS has always struggled to acquire legitimacy in front of other older and more established humanities and social science disciplines in France and to delineate its borders (see Chapter 5, Section 5.5.4 for a summary of the debate on disciplinary frontiers). Also, the balance of power within the interdiscipline is heavily tilted towards communication and media studies. Since the official recognition of this interdiscipline in the 1970s, the place occupied by IS has continued to dwindle. For every professor or student in IS, there are approximately four to five professors and students in communication and media studies. See Ibekwe-SanJuan (2012a, 2012b, 2012c, 2012d) for an in-depth discussion of the effects of this unbalanced interdisciplinary marriage.

1.4.6. The First University Programmes in Information Science

The first higher education programmes in IS suffered from three ills:

(1) the narrow political agenda for STI implemented by the different governments in the first decades of the existence of the field which led to the design of technical and pragmatic courses. The result of this instrumentalisation of IS was that its scholars and students were perceived as the "technicians" of ICS by their communications and media counterpart (see Section 1.4.4 supra);
(2) the rivalry between the professional associations resulted in librarians and archivists not being associated with the academic endeavour of IS and not teaching in the IS programmes in universities. This in turn led to a divide between IS graduates and other more traditional LIS professions (see Section 1.4.2–1.4.3);
(3) the absence of theoretical and historical perspective in the IS education and research which has stalled the development of the field and hindered its visibility in the French higher education landscape.

Fayet-Scribe (1997) retraced the evolution of training programmes in IS in France. The first master's degrees in IS, the master's certificate in Computer Science and literary documentation [*Certificat de maîtrise d'informatique et documentation littéraire*] were created in Grenoble in 1970 and in Bordeaux in 1971. This was no coincidence since Bordeaux was the seat of Robert Escarpit. Bordeaux was also the first and one of the few universities to offer a complete set of university degrees in ICS, from undergraduate to post-graduate and to doctoral studies (Tétu, 2002).

The first IS courses were two-year undergraduate programmes offered by University Institutes of Technology.[39] Students who graduated from the two-year technical undergraduate course either changed orientation, went out to work in the private sector or pursued a master's degree in information documentation. Universities offered either graduate or post-graduate programmes hosted in information and communication departments or in other humanities departments (Literatures, History, Arts, Languages).

[39]Institut Universitaire de Technologie (IUT) equivalent to University technical colleges for vocational training.

With the notable exception of the ENSSIB in Lyon and the INTD in Paris, IS education rarely exists as a distinct specialty and in a distinct department.

Hence, students applying to enter into a master's degree in IS would typically hold a first degree in another field. The result is that IS and documentation are populated by students with very diverse academic backgrounds, ranging from the Sciences (chemistry, biology, physics) to mathematics, economics, arts, social sciences and humanities, most of whom have had only one or two years introductory course in an information- or documentation-related area. As a result, most IS courses are constructed as a beginner course in some specialty. The consequence is that IS is saddled with bits and pieces of courses, each giving a very partial view of the field and resulting in the fact that information professionals graduating from these courses have a narrow area of technical expertise and lack theoretical and historical knowledge about the origins of the field.

Until the year 2000, it was difficult to pursue a complete cycle of degree courses in IS in French universities. The multidisciplinary make-up of the field can foster interdisciplinary research but it has proved to be a stumbling block because scholars and students do not "speak the same language" and do not have the same research traditions.

With the introduction of the Bologna Process[40] in 1999, the French ministry of education put in place reforms to align France's higher education degrees on the three levels: Bachelors–Masters–Doctorate. A handful of French universities now offer a full cycle of training in IS. IUTs are still delivering their two-year undergraduate technical diploma to students who can now complete their IS training with a graduate year in order to obtain a bachelor's degree and then a two-year master's degree. As IS training programmes rarely exist in distinct departments or schools but are embedded in other departments, it is difficult to ascertain how many IS courses are offered in French universities. A recent effort to map European education in LIS[41] may in future provide a more up-to-date picture of IS courses in French higher education institutions.

1.4.7. The Research Landscape of Information Science in its First Decades of Existence

A previous study by Palermiti and Polity (2002) analysed the degree of institutionalisation of IS in France in its first two decades of existence,

[40]More information can be found here: https://en.wikipedia.org/wiki/Bologna_Process.
[41]The map of LIS courses in Europe can be viewed here: http://lis-education-europe.com/.

roughly between 1974 and 1994. They examined the state of its higher education and research apparatus. To measure the state of its research, they looked at the number of doctoral theses defended, the number of departments and research groups, the make-up of students, the scholarly publication channels, the presence and number of dedicated conferences and the existence of publication channels in the field. They found that the fragmented structure of the higher education courses in IS was also mirrored in its research landscape.

In its first two decades of existence, two theses on average were defended per year. This was not altogether surprising given that the field was still in its infancy then but a second reason which accounted for the small number of theses was the small number of full professors in IS who could direct doctoral research.[42] There was a resistance from the other social sciences and humanities disciplines which did not wish to see this new interdiscipline flourish and thus resisted the creation of full professor positions in universities (Palermiti & Polity, 2002). A third reason for the small number was the lack of visibility of IS which resulted in some of its theses being wrongly indexed under "Sciences, techniques and applied mathematics". Two reasons probably account for this: the academic background of the thesis supervisors who came from the sciences, and the lack of appropriate subject headings at the time for documents produced in the new IS field. Also, the main topics of the theses were oriented towards information systems. According to Calenge (2000, p. 21), 45% (135 theses) were preoccupied with automated documentation (*informatique documentaire*) while only 8% of the theses (22) were concerned with the sociology of literacy [*sociologie de la lecture*].

Four Social Science institutions dominated the IS landscape in those first two decades. The EHESS[43] in Paris led by Jean Meyriat and Madeleine Wolff-Terroine produced the most number of theses. Their doctoral students were mostly former INTD graduates, the institution co-founded by Suzanne Briet. The EHESS was very instrumental in setting up the first doctoral courses and seminars in France.

---

[42]In France, there is a clear distinction between the status of Associate Professor (Maître de conferences) and Full Professor (Professeur des Universités). The latter is someone who has defended a further thesis, the Habilitation à Diriger des Recherches (HDR), usually several years after being Associate Professor. This HDR officially gives him/her the right to supervise PhDs. Although in practice, an Associate Professor can supervise a PhD candidate, this must be done in co-direction with a Full Professor or someone who has the *Habilitation* (HDR). This reduces the number of PhDs that an Associate Professor can supervise and also hinders the accession to the status of Full Professor since one of the implicit requirements to qualify as Full Professor is to have acquired experience in directing a PhD.

[43]Ecole des Hautes Etudes en Sciences Sociales (The School of Higher Studies in Social Sciences).

The Pierre Mendès France University in Grenoble came in second place by number of theses defended between the late 1980s and the early 2000s. This group was led by Jacques Rouault, a computer scientist who specialised in automated documentation and Natural Language Processing (NLP). The focus of this research group (where the author of this book prepared her thesis) was on developing and implementing methods based on linguistic and statistics for automatic document indexing and information retrieval (IR). For several years, members of this research group tried to build a large coverage linguistic parser for the French language. The outcome was a prototype that was never fully implemented but which served as the foundation for the design of more targeted language technology applications such as automatic indexing and information retrieval.

The research group in Bordeaux 3 University was led by two of the founding fathers of the discipline – Robert Escarpit and Robert Estivals (1927–2016). The latter specialised on bibliology, a topic on which most of the theses he directed were centred. For Estivals, "bibliology is the science of the written word" and "one of the sciences of information and communication. [...], it includes editology, systemics and political sociology".[44] Estivals co-founded the International Association of Bibliology in Tunis in 1988 over which he presided.

The fourth research group in Lyon was led by Richard Bouché and Philippe Le Guern. This group also worked on NLP, IR and automatic documentation and had for a long time co-directed a common doctoral programme with the Grenoble group. Together, the two groups – Lyon and Grenoble – trained a significant number of second-generation IS scholars who are currently professors in different French universities.

The universities of Aix-Marseille, Paris 8 and Toulouse came in the fifth, sixth and seventh positions respectively in terms of the number of theses produced. The Aix-Marseille group was specialised on bibliometrics and was under the directorship of Henri Dou. The theses defended in Paris 8 and Toulouse were on more heterogeneous topics.

As was the case for graduate and post-graduate courses, students wishing to embark on a doctoral degree in IS came from other disciplines. Most doctorate candidates would be "beginners" in IS research with the result that "there could not really be a progression in the content of the different diplomas and no "accumulation of knowledge" (Palermiti & Polity, 2002). In the rare cases where doctoral students had completed a graduate

[44]Cited in Olivier Le Deuff, Estivals (1987), Robert Estivals, chercheur avant-gardiste en SIC, (Novembre 2016). Retrieved from https://www.reseau-canope.fr/savoirscdi/societe-de-linformation/le-monde-du-livre-et-des-medias/les-penseurs-de-linformation-de-la-documentation-et-de-la-pedagogie/robert-estivals-chercheur-avant-gardiste-en-sic.html.

and post-graduate degree in information-documentation, because of the technological orientation of such courses, they often lacked the necessary historical, theoretical and epistemological background to back up their empirical and applied research. This heightened the impression in the eyes of communication and media colleagues – that IS was the technical and applied arm of the interdiscipline and therefore that its professors were technicians, not scholars.

This perception was further buttressed by the fact that research in IS in its first decades of existence (1980–1990s) was influenced by the academic background of the first generation of professors of whom a significant number came from the sciences. At the time, a wave of scientism was sweeping through the social sciences which led social science scholars to seek alliances with the hard or experimental sciences in their quest for scientific legitimacy. Hence, the first two decades of IS were marked by an alliance with computer science and AI. A lot of hope was placed on the promises of the latter to automate knowledge-intensive activities by building expert systems for translation, knowledge representation and reasoning. Hence, the first-generation IS scholars sought theories and models from linguistics, mathematics and statistics for automatic indexing, IR, automatic translation, expert systems, bibliometrics and scientometrics. Research on man–machine interface and information-seeking behaviour also relied on theories and models from cognitive psychology and AI.

Many of the theses produced in these first two decades were thus rooted in the systems-oriented paradigms. This was especially true of the Grenoble, Lyon and Marseille research teams as well as some research teams in Paris. One can then consider that the first two decades of IS in France marked its "linguistic and AI turn" and coincided on the political level with the narrow political and instrumental policies for STI implemented by successive French governments (Ibekwe-SanJuan, 2012b). However, as AI had failed to live up to its promises, the dream of being able to build large coverage semantic parsers for the analysis of human language has since evaporated. Furthermore, the first generation of IS professors who championed the systems-oriented paradigm and quantitative models have retired and their research themes have gradually disappeared from the IS landscape and have since been taken over now by computer science.

Since the 2000s, a second-generation group of scholars inspired by hermeneutics, socio-constructivism and systemics has sought to anchor the ICS interdiscipline firmly in the social sciences and humanities. As a result, a marked preference is now being shown for research based on subjectivist human-oriented paradigms and on qualitative and reflexive methods. The systems-oriented paradigm is now considered as *persona non grata* and as falling outside the frontiers of the ICS interdiscipline.

More importantly, the historic works of the European pioneers of documentation (Paul Otlet, Henri La Fontaine, Suzanne Briet, Robert Pagès, etc.) on documentation, IS and knowledge organisation systems have been neglected. With the notable exception of the Bordeaux group led by Robert Estivals on bibliology, very little research was done on the history of documents and on public libraries. Although a special track called "Professions of the book" [*Métiers du Livre*] has been created in Information-Communication departments in the Institutes of Technology, it remains a technical two-year programme aimed at training professionals to work in the publishing, cultural and patrimonial institutions such as libraries and museums. The more scholarly study of books and documents has been appropriated by historians, linguists and literary scholars. For instance, Roger Chartier, a historian and professor at the famed "Higher School of Social Sciences" [*Ecole des Hautes Etudes en Sciences Sociales* (EHESS)] who went on to become a member of the French Academy of Science [*College de France*] contributed to the French tradition of research on the history of the book [*histoire du livre*] which in turn inspired research in historical bibliography worldwide.[45]

Since the 2000s, IS in France has returned to its humanistic roots of the late nineteenth and early twentieth centuries. However, contemporary IS scholars do not claim affiliation with Briet and Otlet. There is yet no cohesive body of research themes which can be said to characterise their research. What we have witnessed is an oscillation from one end of the spectrum – from a systems-oriented paradigm favouring computer modelling and empirical research, to the subjectivist-reflexive paradigms favouring qualitative research on the social and cultural dimensions of information and of ICTs. Marcia Bates (1999), the *Grande Dame* of IS in North America observed that a similar oscillation has been taking place within the IS community worldwide, with the pendulum currently swinging back to methods and theories from the social sciences. Her observations about the socio-technical nature of IS is in agreement with those made by Jean-Paul Metzger (2002), a retired IS professor in France about the necessity for IS not to neglect one aspect of its dual nature. Both scholars outlined three research questions that IS must address in order to fulfil its dual nature:

(1) The "physical question", "what are the features and laws of the recorded-information universe?" in Bates (1999) is aligned with the "formalisation and computation" pole of IS in Metzger (2002);

[45]More information on the development of an international research movement on historical bibliography gathered around the "Society for the History of Authorship, Reading and Publishing" (SHARP) can be found here: http://www.sharpweb.org/main/purpose-history/.

(2) The "social question", 'how do people relate to, seek, and use information?' in Bates (1999) corresponds to the "information search behavior" (user studies) pole in Metzger (2002);
(3) The "design question", "How can access to recorded information be made most rapid and effective?" in Bates (1999) can be aligned with two research poles that Metzger (2002) called "information production process" and its "modes of information dissemination".

1.4.8. In Search of Scientific and Cognitive Recognition

There are two ways in which a scientific field can legitimise itself, through cognitive and social institutionalisation (Whitley, 1974[46]).

Cognitive institutionalisation is the degree of consensus reached on the fundamental concepts, research questions and problems, methodologies, relevant results of the field and the basic common knowledge which confers a cognitive identity to this field with regard to others.

Social institutionalisation is the degree to which a field is structured in terms of its membership size, its social identity and internal organisation, its integration into national policies for resource allocation and for accreditation (i.e., existence of own educational programmes, chairs of professorships and faculty recruitment strategies, funding avenues). Typically, fields with a high degree of social institutionalisation would have their own learned societies as well as communication channels (journals, conferences) in which their members publish. The scientific circle can thus clearly be drawn.

Using Whitley's criteria, IS in France has neither achieved social nor cognitive institutionalisation. There is no highly reputed journal in which its community disseminates their research results. Although the journal *Documentaliste – Sciences de l'information* has published articles by IS scholars, it remains fundamentally a professional journal edited by the ADBS – the association of documentalists and information professionals. Moreover, its future remains uncertain. It ceased publication for a while, it has recently been renamed "Information, Data and Documents" [Information, Données & Documents (*I2D*)] and has yet again announced that it will be ceasing publication in 2018.[47] The scientific publisher Hermès-Lavoisier offers IS scholars a few publication avenues in the form of book series and journals such as "Digital Document" [*Document numérique*] and "The Digital Notebooks" [*Les Cahiers du Numérique*]. However,

[46]Whitley (1974).

[47]The announcement can be read here: https://www.adbs.fr/publications-de-ladbs. Accessed on December 24, 2017.

these two journals are not specific to the IS community. They are interdisciplinary and publish a fair amount of research from computer science and linguistics scholars (amongst others).

1.5. The Interdiscipline of Information and Communication Sciences: *Une Exception Française?*

In a series of articles devoted to the situation of ICS in France, Ibekwe-SanJuan (2012a, 2012b, 2012c, 2012d) questioned the claim made by some scholars that the French conception of IS is fundamentally different from the anglophone one, thus the claim of a sort of "french exception" [*exception française*] in IS. Her analysis showed that there is not one unique conception of IS in the anglophone world and that both cultural zones, anglophone and francophone, share similar conceptions of IS when the discourse in the two languages are stripped bare of their surface terminological idiosyncracies. A deep scrutiny of the theoretical assumptions and the methods underlying the research carried in both geographic and cultural zones revealed a remarkable degree of overlap in their foundational authors, choice of theories, models and methods.

However, one distinctive feature of the French IS with regard to its development elsewhere is its institutional merger with communication and media studies right from birth. The cohabitation with communication and media studies has made it more difficult for IS to acquire a clear identity and visibility in the French higher education landscape.

However, after more than half a century of autonomous existence, signs of a *rapprochement* between LIS and communication studies are visible in other parts of the world. In North America, there are increasing mergers of the two fields at institutional and educational levels. For instance, Rutgers University in New Jersey has a School of Information and Communications with three departments: Communication; Journalism & Media studies; Library & Information Science. At the undergraduate level, this school offers an integrated course that combines the three fields, "Digital Communication, Information and Media Minor (DCIM)"[48]. At the master's level however, the tracks are separated. South Carolina University has a "College of Mass communication and Information Studies"[49] which includes the School of Journalism and Mass communications, and the School of Library and Information Science. The University

[48]http://comminfo.rutgers.edu/. Accessed on December 20, 2017.
[49]Website: http://www.sc.edu/cmcis/.

of Western Ontario has a Faculty of Information and Media Studies[50] with separate master's tracks in Media Studies, Journalism and LIS. Kent State University's College of Communication and Information[51] went a step further by offering a doctoral programme in information and communication.[52] The future will tell whether these are real interdisciplinary programmes or an opportunistic merging of several doctoral programmes under one institutional unit.

Outside of North America, the School of Communication Studies of Nanyang Technological University in Singapore which was founded in 1992 created an Information Studies division in 2001. In 2002, the school was renamed School of Communication and Information, and in 2006, it became the Wee Kim Wee School of Communication and Information in homage to the first president of the country who had been a journalist.[53]

These examples are illustrative of strategic mergers at the institutional level, due mainly to budgetary and tactical considerations, rather than an epistemological and ontological fusion of the three fields of LIS, communication and media studies. Furthermore, in the above examples, the mergers took place *a posteriori* of the emergence of the different fields, whereas in the French situation, the fusion of IS and Communication Science into a single interdiscipline entailed a common academic history. Portugal where the field of LIS is struggling for recognition may well follow in the footsteps of the French ICS (see Chapter 4, Section 4.6.3).

[50]Website: http://www.fims.uwo.ca/.

[51]Website: https://www.kent.edu/cci.

[52]Website: https://www.kent.edu/cci/academics/doctoral.

[53]For a history of this institution, see http://www.wkwsci.ntu.edu.sg/aboutus/Pages/Our-History.aspx.

Chapter 2

Emergence of LIS in Yugoslavia: Božo Težak's Information Super-structures

This chapter recalls the crucial contribution of the Yugoslavian pioneer Božo Težak in the emergence and recognition of Library and Information Science (LIS) in Yugoslavia. In sections 2.2 and 2.3, Težak's education and American connections are recalled. In section 2.4, his four LIS super-structures, all created at the university of Zagreb, are described: the International Permanent Exhibition of Publications (ISIP); the Referral Center (RC) of the University of Zagreb; the Center for the Study of Librarianship, Documentation and Information Sciences (CSLDIS); and the journal *Informatologia Yugoslavica* (INYU) (section 2.5). Finally, his legacy to LIS in Yugoslavia and in particular in Croatia is discussed in section 2.6.

2.1. Attempts to set up LIS Education after the Second World War Era

As was the case in other European countries, before the Second World War, librarians in Yugoslavia were trained as apprentices, on the job (Čveljo, 1972, p. 75). The development of library services in the 1950s brought home the fact that a more formal education was needed for library personnel in Yugoslavia.[1] The first of such formal training was set up in

[1]Yugoslavia was a federation consisting of six republics (or states): Bosnia and Herzegovina, Croatia, Macedonia, Montenegro, Serbia (Kosovo was not a republic – but together with Vojvodina was part of Serbia as two autonomous provinces) and Slovenia. Due to the Yugoslavian wars of 1991–1992, the federal state has been divided into separate independent states: Slovenia, Serbia, Croatia, Macedonia, Bosnia and Herzegovina, and Montenegro. However, for historical coherence, we will continue to use Yugoslavia as that was the name of the country in which Božo Težak lived and made his contributions to LIS. Visit https://en.wikipedia.org/wiki/Breakup_of_Yugoslavia for more details.

1961 at the Pedagoška Škola, a teachers' training college in Sarajevo. It was a two-year course (equivalent to an undergraduate course) but it was discontinued in 1969. Similar programmes were developed at the pedagogical colleges of Ljubljana in 1964 and in Rijeka in 1967. An attempt to incorporate library education into the university realm was made in 1962 when the "Chair of librarianship" [*Katedra za bibliotekarstvo*] was set up in 1963 within the Faculty of Philosophy at the University of Belgrade. However, this course was also closed down within two years. Efforts to create university courses in LIS were met with resistance because of widespread scepticism about the pertinence of developing a fully fledged university degree to train librarians. It took the determination and dedication of several distinguished librarians (Eva Verona, Mira Mikačić and Ljerka Markić-Čučuković) and of university professors (Gajo Peleš and particularly of the renowned chemist, Božo Težak) both of the university of Zagreb to make LIS a field of higher education and of scholarly enquiry in Yugoslavia.

2.2. Božo Težak: Chemistry Meets Information Science

While he trained as a chemist, Božo Težak (1907–1980) came early on to understand the importance of scientific information while studying at Imperial College London in the 1930s. However, his work on documentation and IS did not begin until after the Second World War. Several reasons may explain how Težak came to be interested in scientific information at that point in time.

First, a geo-political reason: the Yugoslavian President Josip Broz Tito, who governed Yugoslavia from 1945 until his death on 4 May 1980 (the same year and month that Božo Težak also died!), wanted to strengthen Yugoslavia after Second World War and thus encouraged scientific research. He supported three scientific institutes known for carrying out research on nuclear physics: the Jožef Štefan Institute in Lubljana[2] (Slovenia) which was founded in 1949; the Rudjer Boskovic Institute in Zagreb founded in 1950; and the Vinča Institute in Serbia[3] founded in 1948. These institutes needed an adequate scientific information management system and such a project could only be entrusted to a scientist. Hence, Težak's interest for information and documentation processes coincided with the national interest on the subject. Težak had also displayed his organisational talent in LIS by merging the collections of the two university

[2]https://www.ijs.si/ijsw/More%20about%20the%20Institute%27s%20reactor%20centre
[3]http://www.vin.bg.ac.rs/index.php/en/

libraries in 1946: the library of the Department of physical chemistry and that of the Croatian Chemical Society, both at the University of Zagreb. This gave rise to the Central Chemical Library (CCL) which became the largest reference library in Yugoslavia at the time, open to all users free of charge.

Thirdly, Težak was a well-reputed scholar with an international network. This made it easy for him to interact with scholars elsewhere in Europe and in the United States. He was also an active member of many international learned societies and organisations involved with the circulation of information and the advancement of knowledge. In short, much like Paul Otlet, Težak had humanist and universalist ideals and believed that the free flow of scientific information was a key element for the advancement of humanity.

2.3. Božo Težak's Inspirations: European and American Influences

Težak's ideas on how to develop education programmes, research and services in LIS were inspired not only by his deep knowledge of the relevant literature but also by his international network. He was very active in international organisations in charge of standards and methods for disseminating scientific and technical information (STI). He became a member of the American Documentation Institute[4] in 1953 and of ASLIB[5] the same year (Muljević, 1980). He founded the "Association of Scientific Unions of Yugoslavia" along the lines of the "International Council of Scientific Unions" based in Paris whose task it was to classify scientific disciplines. Težak's assistant, Neva Tudor-Šilović (whom we interviewed) was in charge of international relations and was the Yugoslavian delegate of the FID[6] between 1982 and 2000. In 1953, when Težak was Dean of the Faculty of Sciences at the University of Zagreb, he convened a conference with representatives from 24 research institutions in order to gather ideas on how to build an efficient information system.

---

[4]Ancestor of the current ASIST (www.asist.org)

[5]*Association of Special Libraries and Information Bureaux* founded in 1924 and long-time editor of several LIS journals *ASLIB Proceedings*, *Journal of Documentation*, *Library Hi Tech*, amongst others.

[6]The FID evolved from Paul Otlet and Henri La Fontaine's International Institute of Bibliography (IIB) founded in 1895. The IIB became the Institut International de Documentation (IID) in 1931, and then the FID in 1937. The FID was dissolved in 2002; *Source*: https://en.wikipedia.org/wiki/International_Federation_for_Information_and_Documentation.

A Rockefeller grant in 1951 enabled him to make his first visit to the US where he familiarised himself with current technology for data handling (Dj. Težak, 2010a, 248).

In the winter of 1964, a UNESCO-OECD funding enabled Težak and three of his collaborators to undertake an arduous travel of several weeks which took them to the Federal Republic of Germany, Denmark, Sweden, Holland, England, France, Switzerland and Italy,[7] in order to learn how information science and information technology were being developed. At the time, punched cards and magnetic tapes were the latest technology in information handling.

In 1967, a Fulbright grant took Težak back to the US for an eight-month sojourn during which he visited the country's major infrastructures and centres for scientific information and chemistry research, two of his passions that he managed to combine. Awed by the vast projects for building public and university libraries in the US where access to books and other publications were open to the public, he lamented the very closed attitude to publications which was the norm in his native Yugoslavia:

> In New York, construction will start in the autumn of the foundations of the biggest open library with 2, 000, 000 volumes on open shelves of the New York University. In this country, everything is closed, everything is only stored; we are a real reservation.[8]

During this trip, Težak also discovered Referral Centres (RCs), especially the Library of Congress's Referral Centre[9] which became a model that he tried to replicate on his return to Yugoslavia.

---

[7]A day-by-day account of this trip is given in Muljević (2010, pp. 248–252).

[8]In a private letter to Filip Kneževic dated 1967-05-17 and cited in Djurdjica Težak (2010b, p. 254).

[9]Težak made reference to the role model of this centre in one of his articles published in the first issue of *Informatologia Yugoslavica*, 1969, n°s 1–4, pp. 31–41: "The role of small countries in International Information systems with special respect to Yugoslavia' when he wrote 'The first formation of such type on the national and supranational level is the National Referral Center of Science and Technology at the Library of Congress in Washington" (p. 33). The referral centre is described as the "information desk" of the scientific and technical community. Operating in the Science and Technology Division, Library of Congress, in Washington, DC, "the centre provides a single place to which anyone may turn for advice on where and how to obtain information on specific topics in science and technology" Source : http://onlinelibrary.wiley.com/doi/10.1029/EO053i001p00014-02/abstract?systemMessage=Wiley+Online+Library+will+be+unavailable+on+Saturday+12th+August+at+3%3A00+EDT+%2F+8%3A00+BST+%2F+12%3A30+IST+%2F+15%3A00+SGT+for+4+hours+for+essential+maintenance.+Apologies+for+the+inconvenience

In the summer of 1971, he travelled again to the US, visiting libraries in Washington DC, in New York and in Boston. An American Library Association (ALA) funding enabled him to return yet again to the US in the autumn of 1975. There he attended a meeting of the American Society for Information Science (ASIS), visited and delivered lectures at more than sixteen university libraries in several American cities. He visited Eugene Garfield's Institute for Scientific Information (ISI) and the LIS School at Drexel University in Philadelphia (Dj. Težak, 2010c, p. 262). He also met with Robert Hayes, an IS pioneer[10] at the School of Library and Information Science at the University of California, Los Angeles (UCLA). This led to a long and fruitful cooperation as well as a lasting friendship (Hayes, 2010, p. 348). In his efforts to establish information and documentation as a field of education, professional practice and as an industry in Yugoslavia, Težak was clearly more influenced by American pioneers than by his European counterparts. His early articles contained references to J.C. R Licklider's 1965 book *Libraries of the future* and to Don Swanson's pioneering work on improving communication among scientists.

Težak's knowledge of state-of-the-art information services and techniques in other countries enabled him to develop the Referral Center of the University of Zagreb in order to deal with the exponential growth of information of the 1960s and to replace the old library cataloguing systems with the new technologies of the time – namely microfilms and computers (Težak, 1969c).

2.4. Božo Težak's Information Supra-structures: Influence of Systemics and Cybernetics

Težak's vision for developing state-of-the-art information and documentation systems in Yugoslavia reflected his deep-seated systemics and cybernetic view of information processes in which control was the goal. All of his information-centred initiatives were grandiose in their ambition and all-encompassing in their scope. Four of his projects will serve to illustrate this: the International Permanent Exhibition of Publications (ISIP), the Referral Centre (RC); the Center for the Study of Librarianship, Documentation and Information Sciences (CSLDIS); and the journal *Informatologia Yugoslavia*, all of which were established within the University of Zagreb. The first three were information supra-structures that were intrinsically intertwined and "organically related" (Šercar, 2010, 283).

[10]See also https://www.asist.org/pioneers/robert-m-hayes/

The journal *Informatologia Yugoslavia* (INYU) served as the communication platform for Težak's LIS projects and visions.

2.4.1. The International Permanent Exhibition of Publications

A staunch advocate of interdisciplinarity, Težak was critical of overspecialisation and of the artificial barriers erected between narrow specialties within the same field. He was equally critical of the artificial separation made between "sciences, arts, experiences and interests" (Težak, 1970a, p. 27). He therefore encouraged the melding of the various specialties within the information professions.[11] He set up the "International Permanent Exhibition of Publications" as a non-conventional institution which embodied several functions simultaneously. Težak (1970a, pp. 28–29) himself defined its functions as encompassing that of a "clearing house for publishers [...], a show window for booksellers, a bibliographical center for new publications, a reference library mainly, [...] and above-all, a permanent and changing exhibition of new publications from all over the world".

The permanent exhibition of publications was aimed at enabling people to be "acquainted with contemporary scientific, professional and artistic works, books and journals, and thereby contributing to the development of science and technology" (Bauer, 2010, p. 218). It was a sort of permanent book fair that brought together publishers, booksellers, librarians, documentalists, buyers and users in one place.

Težak was inspired in this project by a contemporary fellow countryman, Dr Josip Štefinović (1888–1958) whose original idea it was to establish a permanent exhibition of publications. A lawyer by profession, Štefinović managed the Science Bookshop in Zagreb between the two world wars (Bauer, 2010, p. 217). In 1950, he began gathering a collection of books which he called the "Permanent Exhibition of Foreign Professional and Vocational Literature". The project became very successful and attracted hundreds of publishers who sent their books to the exhibition free of charge. In 1951 and with Težak's backing, a Senate meeting of the Faculty of Engineering of the University of Zagreb agreed to set up this Permanent Exhibition in the Central Library of this Faculty. However, the initiative raised suspicion from the authorities who could not understand how so many books could arrive from abroad free of charge. Recall that this was just after the Second World War and the beginning of the Cold

[11]Težak quoted a study as identifying 18 profession-oriented and 270 project-oriented services in abstracting and indexing just in the documentation field in the US at that time (Težak, 1970a).

War when everything coming out of Germany (and of the western world in general) was viewed as very suspect in the Eastern bloc. Consequently, Štefinović was placed under investigative custody for two months before the misunderstandings were cleared up and the exhibition was formally authorised to begin functioning (Bauer, 2010, p. 218). In 1953, the Permanent Exhibition became a University institution thus conferring it a more central and permanent status. Težak became its director in 1955 and introduced the use of the Universal Decimal Classification (UDC) system for classifying books and journals (Bauer, 2010, p. 219). In 1957, the University Council adopted Težak's motion to rename this exhibition the "International Permanent Exhibition of Publications" [*Internacionalna Stalna Izložba Publikacija*] which gave rise to the acronym ISIP in Croatian. When Težak obtained from the University Council the creation of the RC in 1967, the ISIP became one of its sub-structures in 1968.

At its highest point, the ISIP attracted submissions of printed materials from thousands of publishers all over the world and collaborated with international organisations such as the Food and Agricultural Organisation (FAO) in Rome, the UNESCO in Paris and the United Nations in New York. The ISIP became an information awareness centre whose importance for countries facing travel restrictions cannot be over-emphasised. During the Cold War and the isolation of the socialist Yugoslavia, generations of scientists remember the ISIP as the only place where they could read "several thousand titles of international scientific journals and books" (Tudjman, 2010, p. 266). Convinced that progress in human society is impossible without the adequate flow of information and in particular of scientific information, Težak linked ISIP's activities in Zagreb with other STI centres across Croatia (Rijeka, Zadar, Split and Osijek), thus enabling other scientists in Yugoslavia to maintain a level of scientific activity during the cold war (Tudjman, 2010, p. 267).

After Težak passed away in 1980, the ISIP continued to be part of the RC and of its successor – the Institute of Information Science and Croatian Information Documentation Referral Agency (HIDRA) – for another 15 years. In 1996, it was integrated into the National University Library of Zagreb as a separate department (Bauer, 2010, p. 222).

Težak's ISIP is rather Otleian in its ambitions *albeit* on a smaller scale. In the same way that Paul Otlet and Henri La Fontaine conceived the *Mundaneum* (Palais Mondial) as a receptacle of the world knowledge, classified according to the UDC, Težak envisaged the ISIP as a place where the world's most important publications will be on permanent exhibition and accessible to all. However, Težak did not go as far as Otlet and La Fontaine in assembling and indexing all existing publications in his time. We also found no explicit reference to Otlet in Težak's writings.

2.4.2. *The Center for the Study of Librarianship, Documentation and Information Sciences*

The second major information institution built by Težak was the Center for the Study of Librarianship, Documentation and Information Sciences (CSLDIS), the educational arm of his many-pronged information initiatives. The CSLDIS was created in 1961 and offered a post-graduate course in the broad field of "Information Sciences". Težak believed that a way to lower the barriers between the different information professionals (librarians, documentalists, archivists, museum curators) was to design an interdisciplinary programme in which these narrow specialisations will be forced to investigate "the physiological, sociological and psychological operations of cognition and communication" (Tudjman, 2010, p. 267). In 1964, the programme was re-organised as an interfaculty and interdisciplinary post-graduate course administered by the Faculties of Philosophy, of Natural Sciences, of Medicine and of Electrical Engineering. The CSLDIS course extended well beyond the realm of Library Science and trained post-graduates in four fields: "i) librarianship, documentation and information sciences, ii) museology, iii) archivistics, iv) information sciences and services (systems engineering, organizational analytics, programming, reprographic, computer and telecommunication methods and techniques, information-documentation- and communication system, computer-assisted instruction, translation services, scientific information system, organization of the scientific research)" Težak (1970b, p. 96).

The interdisciplinary nature of the courses ensured that graduates could work in several information professions but also in different domains (medical, chemical, engineering, etc.). Težak's underlying assumption in designing the CSLDIS programme was that there are common processes underpinning information and communication and their study using knowledge from other fields such as linguistics, mathematics, logic and engineering will elevate the field to the status of a "science" (in the natural science acceptance of the term). Težak was helped in designing and executing the programme by many librarians such as Eva Verona, Mira Mikačić, Tatjana Blažeković, Ljerka Markić Čučuković and the information specialist Tibor Toth. All of them, except for Toth, had PhD degrees. Toth was known for his expertise in retrieving information from databases at the time.

While Težak was a brilliant scientist, he was the only full-time faculty in the CSLDIS programme. Aside from philosophy, there was no collaboration with other humanities and social sciences departments. He therefore sought collaboration from colleagues in the sciences, in particular from the faculty of electrical engineering and computer science. This resulted in an ambitious programme that was oriented towards information systems and information theory.

In terms of organisation, the CSLDIS courses were aligned with those of other European countries and in particular with the programmes in the US. The post-graduate level was organised as a two-year programme. An undergraduate four-year programme on "librarianship and social informatics" was run by the Faculty of Philosophy between 1976 and 1977. From 1984 when a separate Department of Information Sciences was established, the programme was gradually extended to embrace new fields such as museology, archival study, encyclopaedists, lexicology and knowledge management. All these programmes were open to both full-time students and non-residential distance learners. This raised awareness for this field of activity and for the status of librarians in Yugoslavia (Čveljo, 1972, p. 45).

Before the existence of the CSLDIS, IS was not recognised as a discipline. It was impossible to get a doctoral degree in the field anywhere in Yugoslavia and likewise impossible to obtain an academic or teaching position in IS (Prelog, 2010, p. 342). The CSLDIS began admitting doctoral students in 1978. In 1983 (three years after Težak's death), IS was officially recognised as an academic discipline by the Committee for Science, Technology and Informatics of Croatia (Dj. Težak, 2010f, p. 306).

In the same way that the ISIP and the RC were construed as supra-information centres, the CSLDIS was not envisioned as an ordinary university post-graduate institute. Težak endowed it with the functions of a research centre, of a centre for coordinating collaborations between other university units, a publication facility for disseminating research and practice in LIS as well as a bibliographic documentation centre (Težak, 1969d, pp. 69–70).

Such an integration of educational, exhibiting and documentation activities demanded, according to Težak, "the discovery of the most suitable means for harmonious operation of true principles, methods and technologies with a simultaneous solving of fundamental relations between theory and practice" (Muljević, 1980, p. 37).

Težak's vision of the CSLDIS appeared to be that of a knowledge centre embedding other knowledge centres, thus reflecting the influence of the systemics and cybernetics theories on his conception of information-documentation-communication (INDOC) seen by him as a complex system containing sub-systems that can be decomposed into more simpler or equally complex systems, *ad infinitum*.

2.4.3. The Referral Centre: Information Centres within Information Centres

Težak obtained the creation of the RC on 30 September 1967 by decision of the Ruling Council of the University (Težak, 1969d). At the time,

RCs were new and unconventional information services. Težak's achievement in obtaining the creation of a RC in his country is all the more remarkable when one considers that this was accomplished only four years after the Referral Center of the Library of Congress was set up in Washington DC.[12] As he explained in an article presenting the RC:

> Considering documents, information and communication to constitute the essential content of intellectual, educational and instructional and especially scientific activity, as well as the basis for any organization, and having realized the existence and possibilities of very effective new unconventional systems, methods and techniques in emission, transmission, selection and absorption of information, the University of Zagreb has founded the Referral Center of the Universities (further on: RC) to serve, in collaboration with a series of scientific, professional, operational, and economic institutions, as scientific, educational, orientational, completing and integrating institution in the information field in Yugoslavia. (Težak, 1969d)

This passage once again portrays his cybernetic views about the importance of information and communication in all human endeavour. About the scope and aims of his RC, Težak's ambition was again as grandiose as it was all encompassing:

> The purpose of the RC is to act as a body for scientific research as well as a study and operational activity unit primarily within the university, and beyond it as an institution for study, development, recording, orientation, clearing and communication for scientific, professional and operational information sources which are open to the public. The task of the RC is to establish a coherent functional network of documentation, information and communication sharing elements, groups and systems, from traditional and conventional, static ones, in museums, archives and public libraries to dynamic (on-line, time-sharing, real-time) systems for collecting, transmitting, processing and using information in local, regional and global frameworks. (Težak, 1969d, p. 66)

Further in the text, he also conferred academic ambitions on the RC:

> RC is an interdisciplinary, multidisciplinary and transdisciplinary institution for study and operation, primarily at university level but

[12]The Library of Congress's Referral Center began operation on March 4, 1963. Source: Stearns (2009). Accessible at https://www.degruyter.com/view/j/libr.1965.15.issue-1-4/libr.1965.15.1-4.353/libr.1965.15.1-4.353.xml

> with the explicit aspiration to take advantage of already established organizational units, and also to be used by working organizations inside and outside the university. Materials, documents, publications and half-publications have to be, under expedient conditions, made accessible to all interested. (Težak, 1969d, p. 67)

Not only is the RC to be inter-, multi- and trans-disciplinary at the same time, but it aimed also to be a supra-information centre integrating in a single institution, the services of a library, of a documentation centre, of an information centre, of a conference and of exhibition centre on information by integrating ISIP and finally of an educational centre by integrating the courses of the CSLDIS.

The very grandiose and ambitious scope of the RC probably accounted for its break-up into smaller more focused and manageable structures after Težak's death. In 1989, the RC became the "Institute of Information Science". The new name implied that it was dedicated to research in IS; however, the description of its functions showed that it pursued the same missions as the RC (Mihel, 1989) and remained an infrastructural institute for research, training and other activities. The Institute of Information Science was later disbanded in 1995 by the Croatian government (Šercar, 2010, p. 279).

In designing the RC, Težak also aimed to transform the missions of the National and University Library of Zagreb and integrate it into the supra-information institutions he was building. Seizing the opportunity of plans for a new building for the National and University Library, and backed by the University of Zagreb, Težak put forward proposals for the new library to integrate modern technology and techniques, new media and formats including databases and online information, as well as serving broader academic functions. His views are reflected in this passage quoted by one of his collaborators:

> If our networks of university, academy, institutional, industrial, health, agricultural, school, publishing, journalist and other organisations got well connected within the complex of absorption, transmission and emission of special information, we would have a library, an exhibition of publications and an information centre of world significance. To us, this would mean even more; it would be a powerful intellectual (power)station, whose organisation, along with our adequately oriented forces, would surely push us on the entire front of culture, technology, economy and politics into the second half of the 20th century. (Težak cited in Škvorc, 2010, p. 238)

Indeed, Težak always referred to the National and University Library as the "National University Information Centre" (Stipanov, 2010, p. 231). His visions for the new library met with fierce opposition from Matko Rojnic, who was Director of the National and University library for three decades. Rojnic wanted to build a National library in the classical historical sense whose most important mission was the conservation of the national heritage of printed artefacts. Prelog (2010, p. 341), one of Težak's collaborators, described the atmosphere of conflict which pervaded the discussions around the opposing views on the functions and design of the new National and University Library:

> "The early nineteen-seventies were marked by debates on the function and organisation of the future National and University Library in which Težak found himself alone because his proposals of a networked and multimedia library were much ahead of his time as well as beyond the comprehension abilities of the decision makers. Thus, the majority of Težak's ideas and suggestions were met with librarians' hostility. This conflict with the librarians who did not want (or could not) admit "newcomers" or those with broader views into their sphere, was very bitter".

Težak's attempts to have the plans of the new library modified according to his visions were not accepted by a 1970 meeting of the committee charged with the project of the new library. In the end, however, Težak's broad vision of a national information centre rather than a classical library prevailed after his death (Stipanov, 2010, pp. 231–232).

In the same way that Paul Otlet's 1934 celebrated book *Treatise on documentation: the book about the book* [*Traité de documentation: le livre sur le livre*] can be seen as a meta-inquiry on the nature of books while being a book itself, Težak's RC can be said to be about creating "information centers within information centers". The image that comes to mind is that of the Russian *matryoshka* (nesting) doll, each one containing a perfect miniature replica of the same doll, in a spiralist and recursive manner. The RC was the foundational structure and the biggest *matryoshka* doll within which his other LIS institutions were nested: ISIP, CSLDIS and INYU.

Figure 2.1 shows the locations of LIS education programmes in Yugoslavia and the embedded structure of Težak's LIS infrastructures at the University of Zagreb.

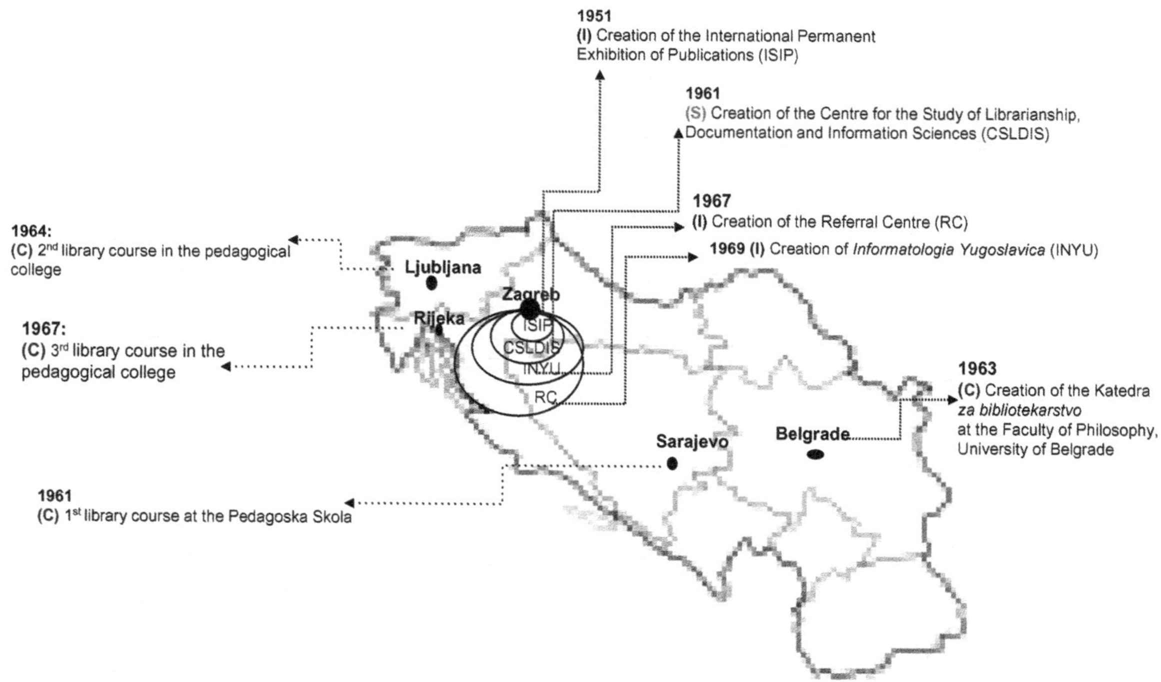

Figure 2.1: Development of LIS institutions in Yugoslavia and Božo Težak's Spiralist and Cybernetic Model of Development of LIS at the University of Zagreb. *Key*: C = Course; D = Diploma/Certificate; S = School; I = Infrastructure.

2.5. *Informatologia Yugoslavica*: A Strategic Communication Tool

Težak's visions and projects to make Yugoslavia an international information hub found a platform in the journal he founded in 1969 which he symbolically named *Informatologia Yugoslavica* (*INYU*) [*Yugoslavia Informatology*], thus reflecting his strong patriotic fibre. The journal was published within the RC and defined as a "semi-publication and a publication of periodic or non periodic character" and "an experimental and working instrument of the Center for the study of Librarianship, Documentation and Information Sciences[13] and also of the International Permanent Exhibitions of Publications, as well as their complement or their common frame, the title of which is the Referral Center of the University of Zagreb" (Težak, 1969a).

The journal was to be open and publish works of academic, professional and practical nature on the "Information Sciences and services". *INYU* not only published scholarly peer-reviewed work in the accepted sense but mostly served as a platform for Težak's scientific, educational and geopolitical visions and initiatives in the information field. Some articles were published simultaneously in Croatian and in English to ensure a maximum dissemination of their contents. About the scientific scope of *INYU*, Težak wrote:

> Collecting, classifying, analysing and scientific processing of data from the domains of museology, archivistics, librarianship, as well as theory and practice of linguistics, logics, mathematics and technology, should be in this journal the clinic material for experimentation and formation of simple and complex information systems. (Težak, 1969a)

The scientific scope of the journal reflected Težak's chemical origins in that he perceived data and information as the basic physical units of empirical experimentation which should lead to simple or more complex systems much in the same way that simple chemical compounds react when assembled to form more complex compounds, giving rise to new compounds and subsequently to new scientific discoveries and theories.

---

[13]Notice the use of the plural form to which we will return later.

2.6. Težak's Ambition for Yugoslavia: An International Hub for Information Science

Težak wanted small countries and in particular his own to be major players in building a networked information society. In one of the first issues of *INYU*, he claimed that:

> The input-output ratio for scientific as well as other material of small countries for their own use is substantially of much higher value than in larger ones. [...] Especially those small countries whose geographical, cultural and political position is outside of the field of direct influence of superpowers, may be very convenient as bridge and clusters for a variety of information centres between various systems, structures and processes of either specialized or general institutions performing the basic information services. (Težak, 1969c, p. 34)

He proposed to build an "Information Center for the Information Sciences and Services" (ICISS) in Yugoslavia as a nodal point, cross-linking international and national networks of special and general information and communication systems. [...] This may be a step toward building up a unified world system of information exchange" (Težak, 1969c, p. 36).

He also claimed that Yugoslavia was in a unique geographic and political position in the world and that information could be a way of bridging the differences arising from its diverse ethnic and linguistic groups. Yugoslavia and Dubrovnik in particular were seen as ideal places for meetings between scientists and students from the East and the West.

In 1977, Težak started a series of international conferences on the theme "Universities on World Information Network" which were held annually at the Interuniversity Center in Dubrovnik. The first edition of the conference was co-chaired with Robert Hayes of University of California in Los Angeles (UCLA). This conference gathered well-known IS scholars from all over the world, including Tefko Saracevic[14] (Rutgers University, US), Eugene Garfield (ISI, US) and Blaise Cronin (initially from Strathclyde, UK before moving to the US). Težak also knew Jesse Shera with whom he had long discussions. Blaise Cronin co-organised conferences with Neva Tudor-Šilović, Težak's assistant in Dubrovnik. Težak started another series of symposia in Zagreb on scientific and professional publications and semi-publications. He proposed the creation of a new journal *Scienta Yugoslavia* that would reflect the application of the latest advances in the field of scientific editing (Šercar, 2010, p. 294). The format of Težak's conferences was informal; there was no

[14]A fellow Croatian who left Yugoslavia for the US where he had a brilliant career in LIS.

official programme and no formal presentation of papers. His scientific events were more of brainstorming sessions with "wide-ranging discussions on the issues involved in the theme Universities on World Information Network" (Hayes, 2010, p. 349). This informal structure proved to be more conducive to serendipitous discoveries and encouraged the cross-fertilisation of ideas. This is reminiscent of the format of the Macy interdisciplinary conferences organised by Norbert Wiener which took place in New York between 1941 and 1960 and resulted in the birth of cybernetics.[15]

Unfortunately, history did not bear out Težak's optimistic view. Small countries did not become the hub of networked information systems and Yugoslavia descended into an ethnic cleansing war in the early 1990s which caused a major and lasting instability in the Balkan States. With the outbreak of the war, it became difficult for people to attend the conferences in Yugoslavia. Dubrovnik was bombarded and the University Center was destroyed. The last conference was organised in 1991 at the onset of the Yugoslavian war. This last conference was also more oriented towards journalism which is why a more LIS-focused series of conferences entitled *Libraries in the Digital Age* (LIDA) was started in 2000 at the University of Zagreb under the double patronage of Tatjana-Aparac-Jelusic and Tefko Saracevic (School of Information, Rutgers University, USA). LIDA later relocated to the University of Osijek and from 2009 to the University of Zadar. LIDA continues to be held every two years and attracts LIS scholars from all over the world. It is often co-chaired with LIS scholars from the United States, thus perpetuating the close collaboration Težak had formed with his American counterparts. From 2016, LIDA is officially co-organised by three institutions: the University of Zadar, the University of Osijek and Rutgers University in the US.

2.6.1. Težak's Legacy to Information Science in Yugoslavia

Težak had a reputation as an honest pioneer in many fields which earned him the trust and support of his peers as well as that of high-ranking officials in the university ruling bodies and in the Ministry of Education. Težak's long-time colleague and collaborator Antun Bauer (1980) praised his "visionary intuition, creativity, energy, uncompromisingness and humanity".

A renowned chemical engineer, information scientist, educator in LIS and innovator, the scope and the ambition of his multiple projects in building information ecosystems were quite ahead of the times and were remarkable given the limited and often rudimentary resources as well as the

---

[15]See the dedicated Wikipedia page at https://en.wikipedia.org/wiki/Macy_conferences.

geo-political situation of his country. He was an influential administrator who held high positions at the University of Zagreb. He was a professor and a one-time Dean of the Faculty of Sciences, Director of the RC and of the CSLDIS. A member of both the British and American Chemical Societies, he was Editor-in-chief of the journal *Croatica Chemica Acta* (CCA) whose international profile he raised. During the period when there a was a ban on importing foreign journals in Yugoslavia, for every copy of *Croatica Chemical Acta* sent to foreign publishers, Težak received in return a free copy of some of the world's prestigious journals as well as bibliographic information. This was a widespread practice internationally during the Cold War era since it enabled people to by-pass prohibitive currency restrictions.

At the time when the Internet was still in its infancy and online access to journals and bibliographic information was a novelty, the Central Chemical library established by Težak in 1946 and the RC created in 1967 became the first libraries in Yugoslavia to receive ISI's *Current Contents* in their print form. The weekly arrival of *Current Contents* was of tremendous importance because it enabled the scientific community of Yugoslavia to stay connected with the rest of the world and to be aware of recent publications in their respective areas of research (Dj. Težak, 2010g, p. 86).

Težak also sought to confer an international status to all his initiatives in LIS by aligning them with state-of-the art technology and practices as well as with international recommendations and standards. In accordance with the UNESCO, the International Scientific Information System (UNISIST) proposed in 1971 by the United Nations was adopted in the RC. As Plenković (2010, p. 319) said of this initiative, "the biggest world information undertaking of that time was presented and implemented in a country that had not yet organised its libraries according to the decimal classification system". In 1979, Težak began discussions with the UNESCO about making the RC of the University of Zagreb a UNESCO centre "entrusted with organising international courses in information handling equipment". The course was to begin in 1981 (Dj. Težak, 2010f, p. 309).

According to his widow, Dj. Težak (2010d, p. 297), "All his introductory speeches when he was opening scientific conferences were about the need of informatisation of scientific and educational systems. [...] Librarians and documentalists were the first to understand his words, and they started the projects aimed at creating our own databases. [...] He published 90 papers in the field of documentation, librarianship and information science, mostly in the journal *Informatologia Yugoslavica*."

Težak suddenly died on the 16th of May 1980, 12 days after the death of President Tito! On the occasion of the 100th anniversary of his birth, a conference was held in Zagreb and a *Festschrift* was published with numerous testimonies about Težak's legacy to LIS (Dj. Težak et al., 2010). There is a consensus amongst scholars in the region that LIS would not have become a scholarly field in Yugoslavia without Težak's efforts:

> It should be noted that before the introduction of Information Science, neither librarianship nor archive nor museum administration had the status of science, they were regarded as services, or at best auxiliary scientific disciplines. (Tudjman, 2010, p. 267)

Čveljo (1972, p. 76) echoed these sentiments:

> The library profession in Yugoslavia is indeed indebted to Professor Težak for his efforts in organizing the study of librarianship in a university setting at a post-graduate level and in initiating training for Documentation and Information Sciences. By integrating Library Science into advanced third-level curricula, alongside the other branches of university studies, Professor Težak has indeed helped to raise the standards of library education and the status of professional librarianship throughout the country.

While Težak's contribution in establishing the scientific infrastructure that helped to develop the field of (L)IS in Yugoslavia is unquestionable, he was not alone in his efforts. Other collaborators, namely a number of librarians and museologists who held MSc and PhD in the field like Eva Verona, Mira Mikačić, Ljerka Markić-Čučuković and I. Bauer, assisted him and contributed to the modernisation of LIS. The first undergraduate programme and academic department in LIS at the Faculty of Philosophy in Zagreb was established thanks to the efforts of librarians, his close collaborators at the CSLDIS.

2.6.2. Težak's Ideals for Free, Open and Networked Information Systems

Težak's ambitions, accomplishments and ideals were not limited to IS or to Yugoslavia. Like Norbert Wiener and the pioneers of the Internet, Težak wanted to build interconnected networks of information systems that would be freely accessible to anyone. In line with the tradition of the twentieth century, he championed openness in science long before open access became the popular movement that it is today, enabled by the Internet and the web. In 1969, he wrote:

> Consequently, we have to think very seriously how to build a multidimensional network spreading over the whole world and enabling us to go on with more or less controlled dialogues between interested individuals and groups including the informational material in various forms of storage. [...] This is the challenge of our time, and not

only in science: How to construct a powerful, efficient and trustworthy system in communication. (Težak, 1969c, pp. 32–34)

Although, Težak was a staunch advocate of the emerging computer technology to enable the interconnection of scientific information institutions and systems, he was also cognisant of the importance of the human element in the whole enterprise. This is shown in his diagram of the functional spheres of documentation (Figure 2.2).

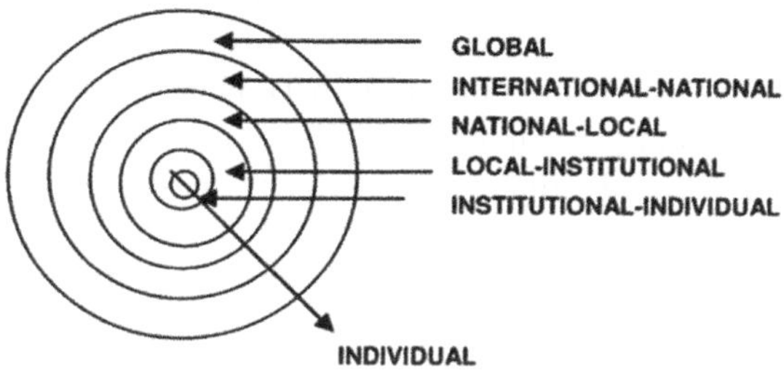

Figure 2.2: Reconstruction of Functional Spheres of Documentation according to Težak's Scheme. *Source*: Dj. Težak (2010h, p. 6).

While the "individual" is the smallest sphere in this diagram, he is also at the centre and the initiator of all the other spheres. However, it appears to us that a spiral instead of spheres would have been more appropriate to illustrate the idea that the different levels are interdependent and not disconnected.

Težak also believed that science was a unifying force that could bridge the gap between societies around the world. Science had therefore to be above reproach because it had the capacity to embody humanity's most idealistic and ethical values. Thus, for Težak (1969c, p. 18), scientific activity should be governed by four principles:

(1) the principle of openness, namely the results must become available to all interested scientists as soon as possible, including the time of experimental procedures and theoretical approaches, since nothing but scientific truth is mandatory (2) the principle of freedom, that is, no one is forced to either publish too soon or told what to publish; the application of these two principles leads to the third: (3) the principle of responsibility for what is published; and the application and reflection of the three principles results in (4) the principle of confidence of the scientific community which will verify the results and then accept them. In general, those involved in fundamental research should fully follow these principles, which constitute the necessary steps towards an open world.

Težak's scientific ethos echo those of Robert Merton (1942[16]), the American functionalist sociologist of science who put forward the four norms or ideals of science: Communism, Universalism, Disinterestedness and Organised Skepticism (CUDOS).

Težak's pioneering work has had a significant impact on the development of LIS in other eastern and central European countries. As Škvorc (2010, p. 242) acknowledged, "from the early nineteen-sixties until his premature demise in 1980, there was no action in Croatia in the field of scientific and professional information that Professor Težak did not take part as an initiator or active participant". His influence was not limited to Yugoslavia as the American scholar Robert Hayes acknowledged: "Even when I became a faculty member at UCLA, in 1964, information science was at most a potential concept for educational programs in the US" (Hayes, 2010, p. 347).

After Težak's death, his collaborators and students from the CSLDIS continued their efforts for the establishment of LIS departments in the country. Currently, LIS departments exist in Ljubljana (Slovenia), Zagreb, Osijek and Zadar (Croatia), Belgrade, Novi Sad (Serbia) and Sarajevo (Bosnia and Herzegovina). Doctoral programmes are offered at the Universities of Ljubljana, Zagreb, Zadar, Belgrade and Sarajevo. However, most students still choose to enrol in Zagreb or Zadar because of their research-based programmes.

[16]Robert Merton, *The Normative Structure of Science* (1942).

Chapter 3

Emergence of LIS in Three Scandinavian Countries: A Tale of Nordic Competition and Cooperation

Of the seven European countries whose history of LIS emergence we recount in this book, the Nordic countries are those for whom there are the most number of publications in English. Nordic scholars have a long tradition of publishing in English. Hence, this chapter relied as much on existing publications as on oral interviews and informal communications with some of the scholars cited in the chapter. First, we retrace the emergence of LIS in Denmark (Section 3.1), followed by its emergence in Sweden (Section 3.2) and in Norway (Section 3.3) before summarising in section Section 3.4 how the tradition of cooperation between Nordic countries after the Second World War encouraged the mobility of LIS scholars from one Nordic country to another. For practical reasons, we were unable to include Finland and Iceland in our study. Finland in particular has a rich and much older tradition of education and research in LIS of which many publications have given an account. The interested reader can consult for instance, Mäkinen, Järvelin, Savolainen and Sormunen (2016) and Mäkinen (2007).

3.1. Copenhagen, the Epicentre of LIS Development in Denmark

Like in many other European countries, before the creation of formal training programmes in the twentieth century, librarians were trained in local libraries much in the same way that carpenters trained their apprentices, that is, "on the job". Concording documentary sources (Harbo 2000, Laura Skouvig[1]) traced the earliest formal library education to the first decade of the twentieth century. From 1910, the library pioneer Andreas Schack Steenberg began offering private courses in librarianship. In 1918, the State made it obligatory for librarians to receive a formal training. A "State School of Librarianship" was thus established in Copenhagen under the auspices of the Government's Consultant on Library Affairs, headed by Steenberg. Initially, this school offered a three-month theoretical librarianship programme followed by an internship at a library. About 15 students enrolled in this programme in September 1918. The first Library Act passed in 1920 saw library education becoming even more formalised and its duration lengthening. The State Public Library Inspectorate was created the same year and entrusted with overseeing the activities of the State School of Librarianship. By 1938, the duration of the librarianship programme had increased to four years and covered both theoretical and practical training. These earliest courses were inspired by the UK and the US models.

3.1.1. The Move from Librarianship to Library and Information Science

When the Royal School of Librarianship (RSL henceforth) was created in Copenhagen in 1956, it was given sole responsibility for training librarians and was placed under the supervision of the Ministry of Cultural Affairs and Communications. In its 1956 act of inception, the RSL was required to not only offer a professional education in librarianship but also to "act in favour of theoretical studies within the different fields of the library system" (Harbo, 2000, p. 5). Preben Kirkegaard, a librarian, became the first Rector of the school and remained in that position until 1983. This research intent was further emphasised by a 1966 law which stated that "the library school shall further research within the library system" (Harbo, 2000, p. 5).

The RSL initially offered educational programme that ran from a few weeks to four years known as the "Basic Library and Information Science Program". This four-year programme was officially not considered the equivalent of a university degree since at the time, the Bachelor of Arts (BA) and the Master of Arts (MA) had not been established at Danish universities. However, it "was ranked equal to the bachelor's degrees from

[1]Private email communication on 6 August 2018.

universities in Denmark" once university degrees became aligned in Europe (Harbo, 2000). The RSL also offered a post-graduate degree in Research Librarianship and Documentation to people who had a prior academic degree in a subject area or worked in academic or research libraries as well as in information and documentation centres (Kajberg, 1988). These candidates could complete their library degree in two years instead of four. The original education programmes of the RSL known as "Section I" were oriented towards training librarians to work in public libraries. From 1969, a second programme called "Section II" was developed to train librarians to work in national research and academic libraries.

Also, in 1969, the RSL started a very small education programme for people already hired as academic librarians in research libraries called "Research librarian and documentalist education" [*Forskningsbibliotekar – og dokumentalistuddannelse* (*FoD*)]. For a long time, the RSL developed continuing education courses for research librarians and documentalists who already had a full academic education and an employment in a research library. These continuing education courses have since ended.

In 1973, a branch of the RSL was created in Aalborg but placed under the same Director as in Copenhagen. The main motivation behind the creation of the Aalborg branch was the co-location with Aalborg University which opened in 1974, and secondly the desire to create a new library education within the university system. However, plans for an integrated educational programme with the University of Aalborg failed for several reasons. First, internally, there were conflicting opinions among the Faculty of the RSL about how this program should be designed. Secondly, collaboration with the University proved difficult given the differences in culture of the governing institutions. Until 2010, the RSL was not under the supervision of the Ministry of Education but under that of the Ministry of Cultural Affairs and Communications. The University of Aalborg on the other hand was under the supervision of the Ministry of Education. The Ministry of Culture fostered a practical "apprentice style" education in the cultural and artistic fields while the Ministry of Education fostered an academic "student-professor" tradition. Being under the supervision of the Ministry of Culture in its first decades protected the RSL from a number of cuts and rationalisations taking place in universities and ensured that library education had a cultural orientation.

The Aalborg branch of the RSL offered only the public libraries education programmes (Section I). There were about 20 teachers at the Aalborg branch, comprised in equal parts of practitioners (librarians) and academics. Until the early 1990s, the policy of the RSL was to hire academics and librarians in equal proportions in its teaching staff. After 2000, the majority of the librarians and the other academics had obtained a PhD. In 1985, the RSL merged its two educational programmes, Sections I and II into one.

The academics at the Aalborg branch leaned towards Marxist ideology and promoted a view of public libraries as being central to social, cultural and political development. In line with the 1970s' social liberation movement, the Aalborg branch of the library school adopted a democratic governance system where every representative of the school, from faculty to secretaries and to students, could vote on the orientation the school should take.

A third law in 1985 emphasised yet again the necessity for the RSL to develop education and research not only on libraries but also on documentation and IS. Hence, "education and research had been placed on an equal footing" (Harbo, 2000). This law also made it possible to create a year-long continuous education programme in 1986–87 in "information systems" for librarians in the private sector. The lectures took place during the weekends and on selected weeks. The programme was headed by Irene Wormell and Peter Ingwersen and eventually led to a two-year Master of Science (MSc) in Information Science and Management in 1990 which Peter Ingwersen headed between 1990 and 1993 (Ingwersen, 2004). The master's programme was aimed at training people for future demands in information management and in information retrieval (IR) systems design. The MSc signalled the expansion of the education programmes at the RSL from training only information professionals to educating information scientists capable of carrying out doctoral research and research work in the information-related disciplines (Ingwersen, 1994, p. 198).

In 1997, the RSL became the Royal School of Library and Information Science (RSLIS), and in 1998, it was granted full university status. Its four-year basic programme was officially recognised as a bachelor's degree (Pors, 2000, p. 12). With the signing of the Bologna Process[2] in 1999, the process of aligning university degrees along the three levels Bachelor, Master and PhD in many European countries began. The impact of the evolution from a vocational education style to full academic style on the professional library community was that gradually the "Librarians' Union changed its member base from that of an association of clerical and professional workers to an association for graduates" (Harbo, 2000, p. 8), thus leading to an upgrade of skilled professionals in the field.

[2]https://en.wikipedia.org/wiki/Bologna_Process.

From 1998, the RSLIS was authorised to set up its own PhD programme of which its first candidate obtained the diploma in 2004 (Audunson, 2005). Before 1998, doctoral studies were done in collaboration with other institutions, in particular with the University of Åbo in Finland (Harbo, 2000, p. 10). The RSLIS was one of the first LIS schools to become a member of the iSchool movement outside of North America.[3]

In the course of its history, the RSLIS underwent many structural and organisational changes. Its 1978–79 annual report showed that it had 21 departments. Departments were responsible for designing their course contents as well as their own research and the heads of departments for hiring his own staff. In the 2000s, it had been reorganised into institutes and had three institutes at one time: "the Institute of Information studies" focused on "Knowledge organization and retrieval." (...) "The institute of Library development" focused "on knowledge management" and "The institute of Culture and media" focused on "cultural policy and the social and political role of libraries, mediation of culture, literature and literary sociology and library history." The RSLIS was "probably the world's biggest LIS-school outside China, with an academic staff of more than 80 and more than 1000 students" (Audunson, 2005).

For several decades, the RSLIS enjoyed a monopoly situation in educating librarians in Denmark (Pors, 2000, p. 14) due to a law which stipulated that heads of public libraries should have a degree from the RSLIS. This monopoly was true for training librarians to work in public libraries but not for the research librarians who were subject specialists trained in universities. However, this monopoly has now been lost as public libraries increasingly employ personnel with other academic qualifications.

From 2013, the RSLIS was absorbed into the University of Copenhagen and housed within its Faculty of Humanities.[4] In 2017, the RSLIS underwent yet another name change and is now known as the "Department of Information Studies" (INF). Jens-EriK Mai, a reputed IS scholar became its new dean. In recent years, LIS has been increasingly marginalised while other humanities fields like cultural studies, media studies, museum studies have grown in prominence. In September 2017, the University of Southern Denmark (SDU) based in Odense began offering a bachelor's degree

[3]The School of Information in Toronto was the first to join outside of the US in 2005, followed by the Berlin school and the RSLIS in 2009. See https://ischools.org/members/ for more details on this coalition.

[4]A history of the RSLIS can also be found here: https://wikivividly.com/wiki/Royal_School_of_Library_and_Information_Science.

programme in "Library Science and Knowledge Communication".[5] This was perhaps not a fortuitous choice given the fact that the RSLIS had a long-term cooperation with this University. According to Harbo (2000), "librarians graduating from the School were admitted to a master's programme covering culture and media" at this university which recognised the four-year basic programme of the RSLIS as the equivalent of a bachelor's degree in Denmark. Figure 3.1 hereafter shows the development of LIS education programmes in Denmark and the central role of Copenhagen.

3.1.2. Competing Research Visions at the RSLIS: The Battle for Pre-eminence

An influx of grants from several national and Scandinavian agencies enabled the RSLIS faculty to develop and maintain a high level of research activity subjected to periodic assessments (Kajberg, 1988, p. 148). The first three decades of research at the RSLIS was focused on public libraries due to the resources devoted to this topic, the opening of the Aalborg branch in 1973 whose educational programme was geared towards public libraries and the creation in 1979 of the journal *Library Work* [*Biblioteksarbejde*] by teachers in this branch. From the 1980s, research broadened to include other topics such as library history, information storage and retrieval, information-seeking, library information and services (Kajberg, 1991, pp. 194–198). Harbo (2000) observed that "As a national institution, the School must cover all aspects of research related to the information sector at large. This is perhaps the real strength of the School: the synergy of minds." At the time of his writing, research activities at the school were organised around "five broad categories, namely information retrieval, knowledge organisation, information management, historical and comparative studies and cultural and media studies".

While inter- and multi-disciplinarity sound good in theory, on the ground it entails the co-existence of diverse research themes and traditions which can stimulate competition and research productivity but invariably engender tensions and conflicts. Whilst the RSLIS was the theatre of research excellence, it was also unavoidably the theatre of power tussles.

The French sociologist Pierre Bourdieu has written a lot about the dynamics of scholarly fields which as he observed are not governed purely by altruistic scientific ideals but like any other social organisation are also

---

[5]Website https://www.sdu.dk/en/uddannelse/bachelor/bibliotekskundskab_og_videnskommunikation.

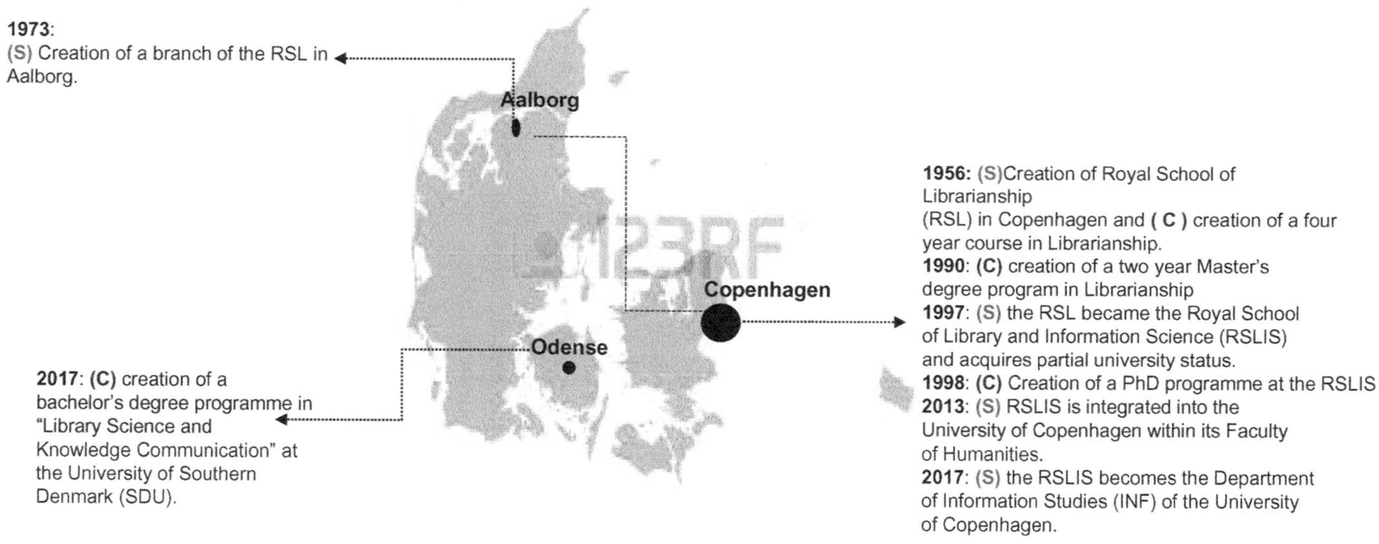

Figure 3.1: Copenhagen, the Epicentre of LIS Development in Denmark. *Key*: **C** = Course; **D** = Diploma/Certificate; **S** = School; **I** = Infrastructure.

governed by a battle for "scientific and social capital" and thus by a struggle for pre-eminence:

> *The* "pure" universe of the most "pure" science is a social field like any other, with its power relationships and its monopolies, its conflicts and strategies, its interests and profits, but where all these invariants take on specific forms.[6] (Bourdieu, 1976, p. 89)

A little further, he reasserts that:

> The scientific field is always a place of more or less unequal battle, between agents endowed unequally with a specific capital, therefore unequally capable of appropriating the products of scientific labour (and in some cases, the external profits of economic or political gratifications[7]) that all the rivals produce by scientific collaboration, by putting into place, all the available means of scientific production. (Bourdieu, 1976, p. 96)

Likewise, the definition of what is "scientific" as opposed to what is not or "less scientific" in any field depends on what the people engaged in this battle define as "scientific" at a given time in history and where the balance of institutional and "scientific" power resides:

> What is always at stake in scientific conflicts in which each of the actors must engage in order to have the value of his products accepted, is the power of imposing the definition of science best conforming to his own individual interests; for the definition of what is at stake is itself part and parcel of the stakes in such a conflict. And the form taken by this struggle over scientific legitimacy depends on the structure of the distribution of the specific capital of scientific recognition among the participants. (Bourdieu, 1976)

Thus, like any social and cultural organisation, academic fields are given to constant debates about their very frontiers or about the scientificity of a

---

[6]In the original French text: "L'univers "pur" de la science la plus "pure" est un champ social comme un autre, avec ses rapports de force et ses monopoles, ses luttes et ses stratégies, ses intérêts et ses profits, mais où tous ces invariants revêtent des formes spécifiques." (Bourdieu, 1976, p. 89).

[7]In the original French text: "Le champ scientific est un lieu de lutte, *plus ou moins inégale*, entre des agents inégalement pourvus de capital spécifique, donc inégalement en mesure de s'approprier le produit du travail scientifique (et aussi en certain cas, les profits externes tels que les gratifications économiques ou proprement politiques), que produisent par *collaboration objective*, l'ensemble des concurrents en mettant en oeuvre l'ensemble des moyens de production scientifique disponibles". (Bourdieu, 1976, p. 96).

given strand of research activity. These debates bring into play not only pure scholarly arguments but also socio-economic, political and cultural considerations. Like other academic institutions all over the world, the RSLIS was subject to internal battles for pre-eminence amongst the different scholars making up its multidisciplinary research traditions.

A first group of scholars gathered around Povl Timmerman favoured a psychological, cognitive approach to LIS. Timmerman was an electrical engineer who worked at the Nuclear Research Station in Risø (Harbo, 2000, p. 7). He joined the RSL in 1971 and led the research on search systems. In a way, he brought "scientific" culture to this humanities institution. Peter Ingwersen who obtained his first degree in librarianship from the RSL in 1973, was recruited by Timmermann to work on the cognitive processes in online human information-seeking activities. After an international research experience at the European Space Agency-Information Retrieval Service in Italy (ESA-IRS), Ingwersen returned to the RSL in 1984 as a tenured Associate Professor. He obtained his PhD from the Copenhagen Business School in 1991. In 1993, he became the head of the Information Retrieval Theory department. Ingwersen garnered international reputation for his high-profile research on the cognitive approach to interactive IR. Other scholars in this group were Søren Kaae, Irene Wormell and Annelise Mark Pejtersen. Søren Kaae was a Danish librarian who worked with the cognitive research group for a short period before returning to practical library work. Irene Wormell worked on information and knowledge management from 1985 till the 1990s. Annelise Mark Pejtersen worked on a search system for fiction. She had access to the biggest computers in the 1970s which were at the only nuclear research centre in Denmark. Ole Harbo, who was for a long time Rector of the RSLIS, also collaborated with Timmerman and Ingwersen in the 1970–1980s on conceptual models to Information Science and on Bibliometrics (Harbo, 2000). When Timmermann suddenly passed away in his fifties, Ingwersen became the torchbearer of this group. This cognitive-oriented group sought to make the RSL a scientific institution, hence, the focus of their research was not libraries but information. Having come to the conclusion that "the expert system approach to Interactive IR (IIR) was too narrow and unattainable", Ingwersen (2004) developed further the cognitive viewpoint in IIR, a theme on which he authored a highly cited article in 1996 as well as the book *The Turn*, co-authored with the Finnish scholar Kalervo Järvelin in 2005. With Irene Wormell and the mathematician and physicist Finn Hjortgaard Christensen, Ingwersen established the Centre for Informetric Studies in 1996 which was headed by Wormell until 2001 when she left to take up a professor position at the Swedish School of Library and Information Science (SSLIS) in Böras. The Centre for Informetric Studies was then closed down by the school's management. When the web emerged as the new research platform in the late 1990s, Ingwersen carried over informetric methods to the web, thereby launching the term 'webometrics' with Thomas C. Almind.[8]

A second research tradition led by Birger Hjørland favoured a subject-oriented approach that emphasised scientific and scholarly communication, domain knowledge and a sociological and epistemological perspective. Before coming to the RSL, Hjørland had studied scientific communication in psychology. In 1973, he was recruited at the RSL to teach in the Department of Reference Work and Documentation, headed at the time by Axel Andersen. Andersen had graduated from the State Library School in 1954 before obtaining a MA in natural history and geography in 1959. He entered the RSL in 1960 where he taught reference literature. In 1996, Andersen published one of the most important books[9] on where to find the right handbook for anything which can be considered the Danish equivalent to Katz (2001).[10] After some years spent working at the Royal Library, Hjørland returned to the RSL in 1990 to head the Department of Humanities and Social Sciences which at the time was the biggest department in the school. His 1995 landmark article with Hanne Albrechtsen on domain knowledge and domain analysis became a new research paradigm in LIS.

A third research tradition around Erland Munch-Petersen and Mette Winge favoured an institutional perspective on libraries. With his background in literary studies and "the sociology of fiction" (Harbo, 2000), Petersen was concerned with the use of bibliography while Mette Winge worked on literary reviews and children's literature. Both scholars contributed to the development of library-oriented literary studies in the 1980s.

A fourth set of scholars at the Aalborg branch of the RSL comprised of Leif Emerek, Anders Ørom and Niels Windfield Lund were influenced in their research and teaching by the philosophical tradition of the "folk culture". The folk culture tradition was founded in Denmark by a nineteenth-century vicar named N.F.S Grundtvig. Local communities and local culture are very important in the Danish society and in Norway to some extent. Denmark is seen as the country of voluntary associations: local initiatives

---

[8]The original publication is Tomas C. Almind and Peter Ingwersen (1997). See also his ASIST Award of merit presentation. Accessible online at https://www.asist.org/about/awards/2016-award-winners/award-of-merit-peter-ingwersen/. Accessed on 29 July 2018.

[9]The title in Danish is "*Veje til viden. Håndbog i informationssøgning*" Copenhagen, Gyldendal, 1996.

[10]See Katz (2001a, 2001b).

emanate from sports association, local peasants or local housewives' associations. These initiatives become the vehicles through which local peasants and country people can be educated, not through formal written education based on books and formal exams but through the "living word", the "passionate speaker", the "conversation" and the "dialogue". The objective of the Folk High School[11] is to foster democracy among the masses and eliminate barriers linked to class structures.

Anders Ørom who was a librarian, shared the same academic background with Emerek: both came from literature studies. Niels W. Lund's background was in history and ethnology. These scholars promoted culture as the central concept of their teaching and research whereas in Copenhagen, information was favoured as the central concept of the discipline. Thus, in a sense, culture was pitted against information as two competing and central concepts of LIS in Denmark.

However, Ørom who worked in the Department of Social Sciences and Humanities also contributed to its research and educational development. Ørom's teaching and research were therefore astride these two research traditions at the RSL.

Other scholars that deserve mention include two sociologists B.V. Elberling and Dorte Skot Hansen. Both worked at the Department of Library Sociology and formed a group. Dorte Skot Hansen headed the Centre for Cultural Policy Studies created at the School in 1998 (Pors, 2000, p. 17).

Lastly, Niels Ole Pors and Carl Gustav Johansen both joined the Aalborg branch of the RSL in 1974 and 1977 respectively and worked on the economic aspect of librarianship (the equivalent of the French *bibliothéconomie*). They were mostly influenced by research on library management in the UK.

Reconciling the competing visions and research traditions of these different groups proved to be a difficult enterprise more so as they were not given equal conditions. For a long time, the proponents of Interactive Information Retrieval (IIR) and "the cognitive view" enjoyed a dominant position.

But competition amongst the diverse research traditions also fostered excellence in the field. A bibliometric study published in 2005 by Persson and Åström found the RSLIS to be the third most-cited LIS institution in the world between 2001 and 2004.

Birger Hjørland and Peter Ingwersen who became its first professors in 2001 have both made significant contributions to the field of LIS worldwide

---

[11]See https://mullsjofolkhogskola.nu/in-english/the-nordic-folk-high-schools/ for more details on this concept.

as recognised by several citation studies (White & Mccain, 1998; Yang & Wang, 2015; Zhao & Strotmann, 2008).

Today, research at the Department for Information Studies (INF) is still as diverse and as multidisciplinary as it was in the past. The department's website features not fewer than eight research groups.[12] The RSLIS has produced LIS scholars of international renown who are currently second-generation Professors in different institutions. Without aiming for exhaustivity, one can mention Pia Borlund, who is professor in the Department of Archivistics, Library and Information of Oslo and Akershus University Colleges (now Oslo Metropolitan University), Marianne Lykke and Birger Larsen, both professors in the Department of Communication and Psychology at Aalborg University, or Jesper Schneider, a senior researcher in Scientometrics at the Danish Centre for Studies in Research and research Policy, Århus University.

3.2. Emergence of LIS in Sweden: a Centrifugal Model of Development

Unlike Denmark, Sweden displayed a centrifugal model of LIS development even if at particular periods, educational programmes were concentrated in a single city. In 1926, a Swedish State Library School (SSLS) was created in Stockholm. Like the Royal school of Librarianship (RSL) in Copenhagen, it was a professional and vocational school whose aim was to train public librarians.[13] For five decades, the SSLS was the only place where a formal education in librarianship for the public libraries could be obtained in Sweden. Academic librarians, on the other hand, were trained "on site", in university libraries until 1972, when the Swedish School of Library and Information Science was set up in Borås.

3.2.1. Loosening the Monopoly of Stockholm

In the 1970s, there was a political desire to decentralise administrations and loosen the central authority of Stockholm in every sector including higher

[12]See list at https://informationstudies.ku.dk/research/research-groups/.

[13]An abridged version of the emergence of LIS in Swedish schools and universities can be found in Margareta Lundberg Rodin's article, "SWEDEN, The librarian's profession is constantly changing. BHS turns 40", *Scandinavian Library Quaterly*, 2012, 45(3), http://slq.nu/?article=volume-45-no-3-2012-5. Accessed on 28 June 2017.

education. This led to the founding of the first proper LIS school, the Swedish School of Library and Information Science (SSLIS) in 1972 in Borås. Until 1997, Sweden had a two-tier system of higher education. The top tier constituted of the major universities endowed with professor positions, graduate programmes and central government funding for research; the second tier was comprised of smaller and middle-sized regional university colleges [*högskolor*] without professor positions or graduate-level studies. The university college of Borås was one such regional university of which the SSLIS became a department in 1977. Designed as a vocational programme, the SSLIS had the task of training librarians for all types of libraries for the entire country. Heated debates about the orientation of its education programme opposed two camps: one camp deemed the programme to be too practical in its bent and not academic enough while an opposing camp (comprised mostly of librarians) deemed the programme to be lacking in practical skills for the job of librarians. A national commission was set up to settle the controversy and it recommended strengthening the academic and theoretic content of the SSLIS programme. As the political scene changed and the social democratic government was replaced by a more conservative coalition, more independence was given to universities and university colleges which had a significant impact on the development of LIS education in Sweden (Lööf, 2000, pp. 36–37).

An important reform in 1991 in the higher education system gave universities the freedom to create their own educational programmes. From 1993, universities were required to re-structure their degrees along the three academic levels: bachelor, master and doctorate. This enabled LIS programmes to align themselves with university degrees in other fields and to integrate broader theoretical and research questions rather than being focused only on practical skills for one type of institution. This also paved the way for the official recognition of LIS as a field of academic enquiry.

The SSLIS in Borås and Gothenburg University began a close collaboration to create a jointly administered new four-year master's programme. The same year (1993), the first doctoral students in LIS were admitted into this programme (Limberg, 2017, p. 3) and the first PhD degree ("fil. dr.") in LIS was awarded to Birger Hjørland from Denmark. In 1998, the SSLIS and the Department of Library and Information Studies of the University of Gothenburg were merged in one department and placed in Borås, thus gathering under one institution, teaching and research activities in LIS (Lööf, 2000, p. 40). This made the Borås-Gothenburg school the biggest LIS school in Sweden offering a complete cycle of education, from undergraduate to post-graduate and research. The SSLIS in Borås managed the educational programmes while the University of Gothenburg received the first Swedish LIS chair, issued the exams for the joint masters and PhDs programmes. From 1995, Borås University College was granted the right to offer master exam in LIS and in 2010, it was granted the right to organise PhD examinations.

By 2000, the SSLIS had four full professors covering four main perspectives on LIS: society and political aspects, knowledge organisation and information retrieval, information behaviour/user studies, and information and library management. One of the professors, Maj Klasson worked on the role of public libraries in local communities and on adult education while Lars Höglund held the professor chair in Gothenburg. As the fame of the SSLIS spread further afield, it began attracting visiting professors from abroad one of whom was Tom Wilson, formerly the Head of the Department of Information Studies at the University of Sheffield. From 1998, Wilson became a professor at the SSLIS, a position he still holds until now (in 2018). In 2013, the SSLIS became the first Swedish LIS school to join the international organisation of iSchools.

From the 1980s, other centres of LIS education emerged in Sweden. At the end of the 1980s, two classes from the SSLIS in Borås were relocated to the University of Umeå. In 1990, the Faculty of Arts at Lund University received an endowment for a professorial chair in "Book and Library History" (Aarek, Järvelin, Kajberg, Klasson, & Vakkari, 1992). This enabled the creation of a LIS programme (BIVIL) headed by Birgitta Olander. Olander had a master's degree in librarianship from the SSLIS school in Borås and a doctorate degree in LIS from the University of Toronto. She worked on "personal information management (PIM) in computer science research and the information behaviour of computer scientists and their scholarly communication".[14] The freedom given to universities to decide their own educational programmes led to Lund and Umeå universities creating their LIS programmes in 1993. The programme at the University of Umeå was housed within the Faculty of Arts and the Department of Mass Media Research and jointly run with Studies of Cultural Administration.

In 1995, Uppsala University created a department of Archives, Libraries and Museum (ALM) headed by Kerstin Rydbeck[15] thus making it the fourth centre of LIS education and research.

The newer LIS programmes in Umeå, Lund and Uppsala were run with very little monetary and human resources. Only through inter-university

[14]A brief bio is available here http://www.kultur.lu.se/en/person/BirgittaOlander/. Accessed on 31 July 2018.

[15]A brief biography of Kerstin Rydbeck is available at http://katalog.uu.se/profile/?id=N94-1225. Accessed on 2 July 2017.

and inter-centre collaboration could they hope to survive and maintain their teaching units and research endeavours with little donations from LIS institutions such as the SSLIS and the Swedish Library Association (Lööf, 2000, p. 39).

Lastly, in 2003, a Department of Library and Information Science was created within the Faculty of Arts and Humanities of Linnaeus University in Växjö and headed by Joacim Hansson, thus making it the fifth and most recent centre of LIS development in Sweden. This department was created in response to the desire of local librarians for a library school with a didactic approach, thus linking LIS to Education Science.

In his analysis of the similarities and differences in the LIS educational programmes offered by Swedish institutions, Audunson (2005) observed that while Borås offered a general LIS education which tried to cover the entire field with its specialties, thus a "broad church" education programme in Blaise Cronin's 1992 typology, the other schools developed "*niche*" programmes targeted to specific areas. Lund university for instance distinguished itself "through extensive use of problem based learning as a pedagogical strategy while the programme in Uppsala had close ties with archives, libraries and museums institutions; Umeå specialised in the field of bibliometric and scientometric research while Växjö stressed "the pedagogical element in librarianship".

To sum up: the Swedish State Library School (SSLS) in Stockholm had the monopoly in librarianship in the first five decades (1926–1972) after which the monopoly shifted to the Swedish School of Library and Information Science (SSLIS) in Borås between 1972 and 1993. From 1993, new centres for LIS education emerged in Sweden due to the possibility given to universities to create their own courses and the move from vocational training in librarianship to LIS education. Collaboration also began between the SSLIS in Borås and the University of Gothenburg; LIS education emerged in Lund and Umeå universities in 1993, in Uppsala in 1995 and in Växjö in 2003. Audunson (2005) surmised that Sweden probably had the highest number of LIS schools "in Scandinavia and probably, relative to the number of inhabitants, one of the highest in the world". All the centres now offer LIS programmes at bachelor, master and PhD levels except for Växjö which does not currently offer a doctoral programme.

As information technology (databases, network, and Internet) penetrate various aspects of society, LIS degree holders faced competition from newer fields such as "information management" or "information brokers" which also produce graduates for the workforce in the private sector. Figure 3.2 hereafter shows the centrifugal model of LIS development in Sweden.

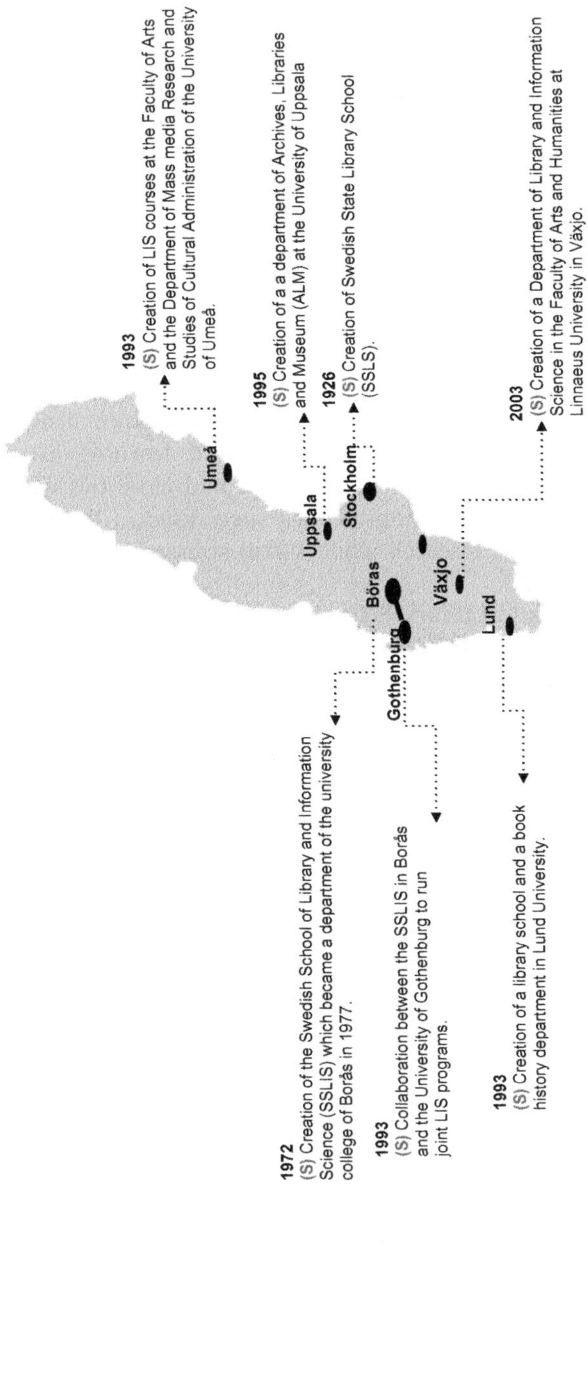

Figure 3.2: The Centrifugal Model of LIS Development in Sweden. *Key*: **C** = Course; **D** = Diploma/Certificate; **S** = School; **I** = Infrastructure.

3.2.2. Dichotomy in Research Poles: Information Science and Library Science

Before the development of LIS degrees in Swedish universities, a few research units emerged in the late 1980s. In 1987, a "Center for Library Research" [*Centrum för Biblioteks – och Informationsvetenskap* (CBI)] was created at the University of Gothenburg by Emin Tengström who also recommended the creation of a chair in LIS in the same university (Lööf, 2000, p. 38). Romulo Enmark[16] who majored in archaeology, arts and ethnology was appointed head of the CBI between 1987 and 1991 while the sociologist Lars Höglund ascended to the professorial chair of LIS in 1993. This was a key development for the discipline because it was the first chair for a professor in the Scandinavian countries. It inspired people to embark on a doctorate research in LIS and attracted candidates not only from Sweden but also from neighbouring countries.

The CBI was partly supported by the Swedish Council for Planning and Coordination of Research which recognised the importance of not only building libraries and training librarians but also of supporting research on libraries. The CBI organised a wide range of activities including international conferences, research seminars, publications amongst which the journal "*Swedish Library Research*" [*Svensk biblioteksforskning*]. After a two-year period as head of the Nordic Council for Scientific Information (Nordinfo), Enmark became the head of the department of SSLIS in Borås in 1993. Between 1997 and 2000, he was the pro vice-chancellor of the University of Borås. In his various management positions, Enmark was a driving force in transforming librarianship into an academic field.

The Borås-Gothenburg SSLIS also founded two research centres: "The Center for Cultural Policy Research and Debate" (CCPRD) which launched the *Nordic Journal of Cultural Policy* [*Nordisk kulturpolitisk tidskrift*], a Scandinavian refereed journal; and the Center for Information Technology Studies that published the journal "*Human-IT*" (Lööf, 2000, p. 40).

A second research pole on IS and bibliometrics coalesced around the Inforsk[17], a research group founded in 1975 within the Department of Sociology at Umeå University and headed by Olle Persson, Anders Lindblad and Lars Höglund. As mentioned above, Lars Höglund was the first LIS professor at the University of Gothenburg and one of the leading figures of Inforsk. The Inforsk group was later renamed "Science Studies"

---

[16]According to his Wikipedia stub https://sv.wikipedia.org/wiki/Romulo_Enmark.

[17]'Forsk' means research in Swedish, Danish and Norwegeian. More information on this initiative can be found at http://archive.is/20130713192025/www8.umu.se/inforsk/. Accessed on 1 July 2017.

although its focus is still on bibliometrics. It is now headed by Rickard Danell[18], a former student of Olle Persson.

In 1983, a Library and Information Science Laboratory (Liblab[19]) was created at the University of Linköping in the Department of Computer and Information Science by a group of scholars of whom the most prominent was Roland Hjerppe.[20] Liblab's research was on information access, document description and representation, thus oriented towards systems design not only for libraries but also for museums and archives. Aarek et al. (1992, p. 35) performed a content analysis of the research output of four Nordic countries (Denmark, Finland, Norway and Sweden) between 1965 and 1989. Their analysis showed that Swedish LIS research was less visible than it could have been during that period because of the low number of doctoral dissertations defended and also because first generation LIS professors published in journals and conferences situated in their "home" disciplines, outside of LIS. This is not without recalling the situation of LIS in France in its first two decades of existence and even till today (2018). As an academic field, LIS still lacks visibility and legitimacy in France.

3.3. Norway: A Terrain of Nordic Collaboration and Innovation in LIS

Norway appears to have been a terrain of innovation where the Danish and the Swedish models of LIS development were brought together to create a Norwegian blend of sameness yet distinctiveness. For interested readers who understand Norwegian, Audunson and Gjestrum (2015) wrote a comprehensive history of LIS education in Norway. For those who do not, we propose below our own blend of the Norwegian history of LIS resulting from both oral and documentary sources.

Before 1940, a significant proportion of Norwegian librarians were trained in the US. Two pioneering figures, Hans Tambs Lyche and Haakon Nyhuus led the "new library order in Norway" at the end of the nineteenth century (Danton, 1957, p. vi) with the former preparing the ground for the latter and for the public library revolution.

---

[18]For more information on Rickard Danell, see http://www.soc.umu.se/english/about-the-department/staff/?languageId=1&uid=rida0001&guiseId=256515&orgId=cf2c8c27071677417e711d8-da163e2dbe94fa2e2&name=Rickard%20Danell. Accessed on 1 July 2017.

[19]For more information, see Roland Hjerppe (1996), Research programme for LIBLAB. Online at http://www.ida.liu.se/labs/liblab/research/2ndResPgm.html. Accessed on 2 July 2017.

[20]A biography of Roland Hjerppe is available at http://liu-se.academia.edu/RolandHjerppe/CurriculumVitae. Accessed on 2 July 2017.

With the outbreak of Second World War in 1939 which disrupted international travels, a Norwegian School of Librarians (NSOL) was founded in Oslo on January 1, 1940. However, due to the German occupation of Norway in April of 1940, students were sent home in May 1940.

Like its counterparts in Denmark and in Sweden, the NSOL was a vocational school that offered a nine-month training course to people with prior experience of work in libraries. Its creation signalled the beginning of structured and formal education in librarianship.

By the early 1970s, the duration of the NSOL course had been increased to three years while incorporating some academic content. Candidates completing this programme obtained a diploma in librarianship. By 1983, the training had been extended to include a two-year programme with an emphasis on knowledge organisation and Information and Communication Technology (ICT). Between 1983 and 1994, candidates who had completed this two-year programme were awarded the title "diploma librarian" [*diplombibliotekar*]. In 1994, the NSOL was integrated into Oslo and Akershus University College which led to it obtaining a professor position occupied by Ragnar Andreas Audunson,[21] a political science scholar. From 1994 till 1997, the title was changed to "majoring candidate" [*hovedfagskandidat*] "Hovedfag" is the Norwegian equivalent of a master's degree after the Bologna process was put in place, although the "hovedfag" was a more comprehensive programme with, among other things, a bigger thesis than in the current master's degree. From 2003 when Norway implemented the Bologna process, NSOL aligned its LIS degrees alongside the bachelor's (BA), master's MA and doctoral (PhD) levels.

3.3.1. Training the "Complete Librarian": An Integrative Model of LIS Education

Audunson and Gjestrum (2012) observed that the goal of the NSOL was "to train "the complete librarian", a librarian who is in command of the entire spectrum of library and information science as a professional field. [...] The purpose is to train professionals who can activate knowledge from all these fields in an integrated manner to respond to the users' need for information." Hence, the first library courses in Norway were of the "broad church" approach as the ones developed at the RSLIS in Copenhagen and at the SSLIS in Borås-Gothenburg.

[21]This part of our historical account on the NSOL relied on Ragnar Andreas Audunson and Liv Gjestrum (2012), as well as on private email exchanges with Ragnar Andreas.

The need for a broad educational approach to LIS is easily explained by the geography of the country: "Norway is a long and narrow, sparsely populated country. Many libraries have only one librarian, and it is essential to possess a broad set of skills" (Audunson & Gjestrum,[22] 2012).

The Oslo University College could not immediately offer a PhD programme in LIS as university colleges were not allowed to do so unless they had more professors and more research programmes. This became possible only in 2013, five years after the first doctoral programme in LIS was established at the Documentation Institute (*Dokvit*) of Tromsø university (see Section 3.3.2 infra). Oslo and Akershus University college of Applied Science where the NSOL programme is located obtained full university status only on the 12th of January 2018 and is now called Oslo Metropolitan University (OsloMet[23]). Its LIS unit, the department of Archivistics, Library and Information Science is also a member of the iSchool movement.

In a similar pattern with Sweden, the monopoly of the NSOL in training librarians was gradually broken through collaboration with universities. From 1995, other LIS centres emerged in Trondheim and in Tromsø. The National University for Technology and the Natural Sciences (NTNU[24]) in Trondheim began offering master's and doctoral programmes in information resources management. Although these courses had a library component, they were more computer-oriented and did not aim to train librarians. However, Ingeborg Solvberg, a professor in these programmes, worked on library systems and digital libraries which were of relevance to the field.

3.3.2. The Documentation Science Institute at Tromsø: A Terrain of Innovation

Unlike the other LIS training programmes in Denmark, Sweden and Norway that focused on library and later on IS, the choice was made in Tromsø to place documentation at the centre of its training and research. Two reasons account for this peculiarity. The first is a political and legal requirement. In 1989, the Norwegian government voted a legal deposit act

[22]At the time of publication of this article, Audunson and Liv Gjestrum were respectively Professor and head of department of the LIS programme at Oslo and Akershus University College of Applied Sciences.

[23]More information can be found here: https://www.hioa.no/eng/HiOA-is-now-known-as-OsloMet-Oslo-Metropolitan-University

[24]The Norwegian University of Science and Technology (NTNU) as it is known now is one of the oldest universities in Norway. Its origins date back to "1767 and the Royal Norwegian Society of Sciences and Letters (DKNVS)". Initially focused on Engineering and Natural Sciences, it now offers diplomas in all the disciplines, including Social Sciences and Humanities. More details can be found at https://www.ntnu.edu/about-research-and-education-at-ntnu.

requiring all published materials, irrespective of their medium, to be deposited at the National Library of Norway which itself was established the same year (Lund, 2009). The National Library which had been dealing only with documents in print was therefore faced with the challenge of handling documents in all kinds of formats (print, audio, video and digital including websites) and from institutions other than libraries (museums, archives) which it had to catalogue, preserve and search. This brought to the fore the need for more innovative and sophisticated techniques and for theories on documents which were not dealt with in the traditional library programmes.

Secondly, as Audunson and Gjestrum (2012) observed, geography played an important role in the way LIS training programmes developed in Norway. The Arctic region of the north had difficulty attracting and retaining workers in every sector with the result that there was a dearth of librarians in the local communities.

The Norwegian Library Association mandated the University College of Tromsø with developing a new education programme to meet these challenges. A local committee was set up to decide how best to design this new programme. One of the questions raised was whether to establish an entirely independent programme in Tromsø or to partner with existing programmes at the NSOL such that teachers from Oslo can go up to Tromsø and teach.

Realising the challenges posed by the new digital media,[25] "the Norwegian committee in Tromsø proposed choosing the notion of document instead of information as the basic concept for the program" (Buckland & Lund, 2013, p. 46). The committee published its proposal in 1989.

The next task was to find someone to design and chair this new programme. In line with the spirit of Nordic cooperation and emulation, a group of Norwegian librarians visited the RSL in Copenhagen in 1987. Niels Windfield Lund who had been an associate professor at the Aalborg branch of the RSL between 1975 and 1988 went on his first trip to Tromsø in 1990. There, he met with some of the people involved in this project for a new LIS programme.

In 1991, Lund wrote a critique[26] of the LIS approach in the Nordic countries which he considered to be dominated by the cognitive paradigm. By so doing, he positioned himself in the Norwegian discussion because the article was published just before a conference in Norway where the future of LIS education in the north of Norway was discussed.[27] Lund considered

---

[25]Recall that in 1989, Internet was still in its infancy and that we were at the dawn of the web.

[26]Niels W. Lund, 'The Library - a world on its own or [...]? National Library Office 1991, ISBN: 82-7195-097-5.

[27]This happened during the Northern Norwegian Library Conference in Harstad Norway in June 1991.

that a document approach offered the opportunity to work not only in libraries but also in museums and archives and on different types of artefacts. The Tromsø project offered him an opportunity to develop his broad conception of documentation.

In 1992, the University of Tromsø tendered a two-year contract for a consultant to develop the new educational programme to which Niels W. Lund applied. The contract was awarded to Ingeborg Motzfeldt, a librarian from Oslo who had been working at the National Library Office. She proposed a traditional library programme that did not integrate the document approach recommended by the 1989 committee tasked with formulating a new programme for LIS in the north of Norway.

In 1993, while Niels W. Lund was a visiting scholar in Tromsø, he undertook a small research project on the role of libraries in small communities which involved going to live in small villages and studying small libraries in the north of Norway. In November 1993, while Ingeborg Motzfeldt organised a conference to propose her LIS education programme, Niels W. Lund launched a new journal entitled "*Norwegian Journal for Library Research*" [*Norsk Tidsskrift for Bibliotek Forskning*].

In December 1993, the traditional programme proposed by Ingeborg Motzfeldt was rejected by the School of Social Sciences at the University of Tromsø because it found the theoretical foundations of the programme to be too weak, thus making the prospects for a university library education uncertain.

In 1994, Niels W. Lund organised a small conference in Tromsø on how to create and maintain a small library programme both in research and education. The first issue of his *Norwegian Journal of Library Research* was published during this conference. The conference was attended by people from the Norwegian Research Council (NRC) sent there to find out how to set up a national research programme in LIS.

Through 1994, there was a dialogue about resuscitating the innovative programme in documentation which the 1989 local committee had recommended. Finally, the University of Tromsø opted for a programme in Documentation Studies. In early 1995, Niels W. Lund was asked to create this new programme. Given this choice of crafting a distinctive new programme on documentation, the Tromsø educational model became independent from the traditional Oslo librarianship programme. Furthermore, it was also decided that the new documentation programme would be lodged inside the University and not in a special vocational school.

By January 1996, Niels W. Lund had created the Documentation Studies [*Dokumentasjonsvitenskap* (*Dokvit*)] programme which had its own department in the University of Tromsø. He became its first professor and headed the unit between 1996 and 1999.

Dokvit was structured as a three-year programme. The first year was a general course in documentation. In the second year, students could specialise either in library or archival studies and in the third year, they could choose a subject matter area (French, history, literary studies, chemistry, etc.) on which to apply the knowledge acquired in the first two years. In 1998, *Dokvit* became the first programme to offer a doctoral programme in LIS in Norway. Hence, within two years of its existence, this new programme was able to offer the full range of university degrees in LIS. Its rapid expansion was due to the fact that the students wanted the full cycle of training in LIS which they obtained from the university governing bodies and from the Ministry of higher Education. Unsurprisingly, this caused some tensions with the older Oslo and Akershus University College. The *Dokvit* programme was a subject of both curiosity and scepticism from the LIS community in Norway as well as in the other Nordic countries. The professionally trained librarians were very sceptical about its ability to train well-qualified librarians capable of cataloguing properly. At the first meeting of the heads of library schools in Nordic countries which took place in Iceland in June 1996, Niels W. Lund had to defend his choice of "documentation" rather than the more established and traditional Library Science or the emerging concept of "Library and Information Science" (LIS). For more details of the *Dokvit* adventure, the interested reader can consult Lund (2007).

While the "distinctiveness" of the *Dokvit* programme at Tromsø singled it out in the LIS landscape in the Nordic countries, the Oslo school remained a dominant force in the LIS landscape in Norway both on the educational and research arena. Between 1996 and 2001, the NRC funded a Library Research programme aimed at developing doctoral research of which the coordination was given to the NSOL. Likewise, the majority of scholarships holders were from the NSOL. From 2012, the NSOL received accreditation to run their own doctoral programme in LIS.

The two major LIS schools, NSOL in Oslo and *Dokvit* in Tromsø have learned to cultivate an "*entente cordiale*" (cordial relationship). They have largely worked together to establish LIS as an academic field in Norway, first, via the *Journal of Library Research* and secondly, via several research programmes. Between 2007 and 2012, both schools were involved in the PLACE research project funded by the NRC and coordinated by the NSOL. Both schools are involved in a second research project till 2020, also funded by the NRC.

Figure 3.3 hereafter shows the dates and places where LIS education developed in Norway.

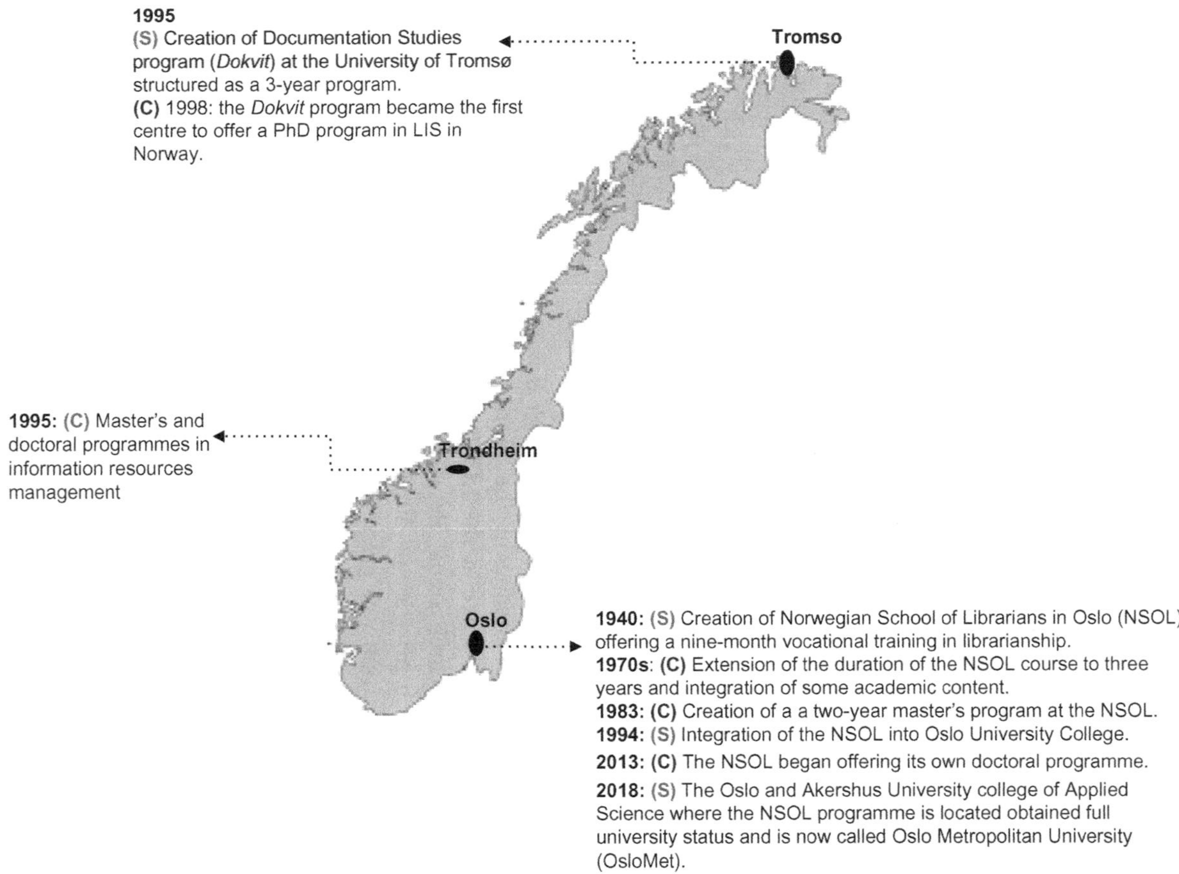

Figure 3.3: The Centrifugal Model of LIS Development in Norway. *Key*: **C** = Course; **D** = Diploma/Certificate; **S** = School; **I** = Infrastructure.

3.3.3. The Power of Serendipity: Coffee-break Meeting and the Birth of the Neo-documentation Movement

In contrast to the scepticism the *Dokvit* programme faced on its home ground, N.W. Lund found much needed support from unexpected and foreign quarters. While he was attending the second "Conceptions of Library and Information Science" (COLIS)[28] conference in Copenhagen in 1996, he met the two anglophone scholars, Michael Buckland and Boyd Rayward during a coffee break. This accidental meeting was to prove significant for the recognition of his documentation studies programme in Tromsø.

Through this meeting, Buckland and Rayward introduced Lund to Paul Otlet and Suzanne Briet's writings which provided him with the much needed historical and theoretical background on documentation theories which his Scandinavian detractors said were lacking in his approach. He also learnt of Rayward's article "Restructuring and mobilising information in documents: a historical perspective" published in the proceedings of the first COLIS conference in Tampere, and later of Buckland's 1997 article "What is a 'Document'?" Having received this crucial encouragement from these two eminent scholars who assured him that documentation theory was "the future", a much bolstered N.W. Lund was able to contemplate the future of his *Dokvit* programme with equanimity and determination.

A discussion on the notion of document or documentation is outside the scope of this book. We will only recall the definitions from the scholars concerned. Lund (2009, p. 41) suggested "to view the document from three complementary angles: a physical, a social and a mental angle, all together enabling a complete description (Lund, 2004). This means that the document does not partly carry each of these three features, but that it is 100% a physical phenomenon, 100% a social phenomenon, and 100% a mental phenomenon. Any document is a physical object as well as a social and mental object. The core issue is how these dimensions are interacting with each other in different ways. This depends on the physical, social and mental environments in which documentation takes place, no matter whether it is in a library, in health

[28]Conceptions of Library and Information Science is a cycle of conferences started in 1991 in Tampere Finland in 1991 by Blaise Cronin (an Irish-American LIS scholar based in the US) & Pertti Vakkari (a Finnish LIS scholar). A history of COLIS conferences can be found here by http://iva.ku.dk/english/research/colis_8/history-of-the-colis-conferences/. Permalink: http://www.webcitation.org/70vnxpaWF.

care, the Arts, business, local as well as virtual subcultures, politics and so forth." For Buckland and Lund (2013, p. 46), the "notion of document [...] was understood to denote, potentially, any signifying thing. *Documentation* was seen as both the process of *documenting* and the outcome of that process".

This wide conception was reminiscent Briet's semiotic conception of documents published in her 1951 manifesto *Qu'est que la documentation?* It is noteworthy that Lund designed his *Dokvit* programme and defended a conception of document similar to those of Paul Otlet and to Suzanne Briet's before he encountered the writings of these two eminent pioneers. A possible explanation may lie in his background. Lund was not a newcomer to the francophone academic scene. He had studied in Paris in the French "Annals" tradition of historiography under Emmanuel Le Roy Ladurie, a notable French historian who became General administrator of the French National Library [*Bibliothèque National de France* (BNF)] between 1987 and 1994. The "Annals" school was founded in the 1920s by Lucien Febvre and Marc Bloch and was so named for their journal "*Annals of economic and social history*" [*Annales d'histoire economique et sociale*]. This school adopted a holistic and interdisciplinary approach to history with much attention paid to details. In the 1960s, it evolved towards a cultural and anthropological approach to history termed "New History" [*Nouvelle Histoire*].[29] The ideas developed by the historians of the Annal's school in the 1930s where Lund trained underscored his broad and semiotic conception of the document.

This semiotic conception of document which underpinned the *Dokvit* programme meant that documents were studied within their contexts of usage and with regard to their users, thus from a holistic perspective combining the three aspects: technical, social and the cultural. The *Dokvit* programme was also guided by two principles: it was constructivist in its epistemological approach and adopted the "problem-based learning (PBL)" in its pedagogical approach. For instance, recommended readings and the projects undertaken by the students were not fixed *a priori* but evolved according to interactions between the students and their instructors. Students and instructors came from multidisciplinary backgrounds.

An echo of this holistic conception of documents as being at the same time physical, social and mental entity resurfaced in the work of a multidisciplinary group of researchers called Pédauque, created in France in 2003

[29]More details can be found on the dedicated Wikipedia pages: https://fr.wikipedia.org/wiki/%C3%89cole_des_Annales and https://en.wikipedia.org/wiki/Annales_school.

to study the evolution of digital documents.[30] However, the Pédauque group was only concerned with digital documents. Lund who was associated with this group considered this restriction to be problematic because digital documents are only a subset of documents that are connected to other kinds of documents.

The coffee break meeting of Niels W. Lund, Boyd Rayward and Michael Buckland at the COLIS conference of 1996 led to a long and fruitful international collaboration as well as to a revival of the documentation movement at a time when information technology and IS were taking centre stage in the field.

Through a series of mobility programmes, the *Dokvit* programme acquired international status. Boyd Rayward came as a visiting professor to Tromsø in 1997. Vesa Suominen, a philosophy oriented LIS scholar from Oulo (Finland) also came as a visiting scholar to the Documentation Institute.

Niels W. Lund himself was a frequent visiting professor to the School of Information Management and Systems (SIMS[31]), University of California, Berkeley, where he was received by Michael Buckland, first in 1999 for a short trip, then in the spring of 2001 for half a year and again in 2005 for a year where he taught document theory to students. In 2003, Lund & Buckland founded a neo-documentation movement called the Document Academy "as an international forum for examining what a document is and how documents can be created, managed, and used" (Buckland & Lund, 2013, p. 46). The first conference of the Document Academy was held at University of Berkeley in 2003.[32]

From 1999, institutional reforms in Norwegian universities led to creating large departments. The Documentation Institute housing the *Dokvit* programme was thus integrated into the Department of Language and

---

[30]Pédauque is the pseudonym chosen to designate a collective group of multidisciplinary scholars in France, created in 2003 thanks to funding by the National Centre for Scientific Research (CNRS) and led by Jean-Michel Saläun. This group was tasked with elaborating research directions on digital documents. It gathered scholars from disciplines such as Linguistics, Computer Science, Semiology, Information and Communication Sciences or History. More information on its history can be found here (in french) http://base.d-p-h.info/fr/fiches/dph/fiche-dph-8220.html. Accessed on 7 July 2017.

[31]Originally founded in the 1920s as the "School of Librarianship" at UC Berkeley, it became the "School of Library and Information Studies" (SLIS) in 1976, it was renamed the "School of Information Management and Systems" (SIMS) in 1994 and renamed again in 2006 the "School of Information" in line with the iSchools movement.

[32]More information on Document Academy can be found here http://documentacademy.org/?index. Accessed on 3 July 2017.

Culture[33] in the Arctic University of Norway (Tromsø) and from 2013, the programme became known as the "Media and Documentation Studies" in order to widen the students' recruitment base.

3.4. The Famed Nordic Cooperation Operating in LIS

Since the Middle Ages, Nordic European countries have had a close connection socially, culturally and politically, sometimes forming political unions, at other times and like all healthy siblings, being at war with each other. These countries share similar languages, folk customs and beliefs, thus a sense of unity. According to Danton (1957, p. 1), "until 1814, when Norway was separated from Denmark, intellectual life was centred in Copenhagen to which most Norwegians went for their higher education". The Folk high school movement originating from Denmark became in the late nineteenth century the symbol of Nordic collaboration.[34]

After the Second World War, there was a political desire to rebuild a strong Nordic identity. Several Nordic organisations were set up to encourage cooperation in all areas, including science, education, arts and culture. It became possible for a citizen of one Nordic country to move freely, live and work in the other Nordic countries. Geographic mobility was also facilitated by linguistic proximity since the three languages, Swedish, Danish and Norwegian, are quite close; one can speak in one of the languages and be understood in the other two.

3.4.1. Getting the Best Players in the Field

In his 2005 article, Ragnar Audunson asked if there was a Nordic perspective in LIS education. He looked at the way LIS education developed in five Nordic countries (Finland, Iceland, Norway, Denmark and Sweden). He concluded that apart from Finland whose LIS tradition programmes date a century earlier, LIS education was vocational at first in the other four Nordic countries; it was aimed primarily at training librarians. The curricula consisted of traditional librarianship subjects such as "Classification, cataloguing and retrieval, the management of libraries and also literature and the mediation of culture and literature". He observed

[33]More information can be found here: https://en.uit.no/om/enhet/forsiden?p_dimension_id=210121. Accessed on 7 July 2017.

[34]https://mullsjofolkhogskola.nu/in-english/the-nordic-folk-high-schools/.

that the earliest LIS institutions were colleges or schools outside of universities and the national system of academic degrees. These schools had a strong literary and cultural component in their curricula which according to him "maybe one of the characteristics of Nordic LIS". He also observed that there was for a long time a monopoly situation where LIS education was concentrated in one centre in the three major countries, in Copenhagen for Denmark, in Borås for Sweden and in Oslo for Norway with Copenhagen maintaining that monopoly while in the other two countries, other centres appeared to challenge it. While our findings largely concur with these observations made more than a decade ago, we found that these features are not the most distinctive of LIS in the Nordic countries since they are also shared by the other Latin European countries we studied (France, Spain and Portugal).

What appeared to better characterise the development of LIS in the three Nordic countries compared to the other four countries we studied is their models of territorial development and their strong tradition of cooperation and competition.

Concerning their territorial development, the Danish model remained resolutely centripetal with everything converging towards a unique centre – Copenhagen, the Aalborg branch being a secondary one. Although the University of Southern Denmark (SDU) in Odense began offering a bachelor's degree in "Library Science and Knowledge Communication" in 2017, the future will tell if this new centre will consolidate its offer away from the centripetal forces of Copenhagen or become one of its satellites. Sweden and Norway on the other hand displayed a centrifugal model of LIS development. Although the first LIS institutions were understandably founded in Stockholm and Oslo respectively, other important centres emerged to challenge the hegemony of the two capital cities.

The second characteristic trait that is a hallmark of the Nordic countries is their strong historic culture of cooperation in science, culture and education. In the LIS sector, joint research projects were encouraged through associations like Norddok and Nordinfo.[35] In 1979, teachers at the Aalborg branch created the journal *Library Work* [*Biblioteksarbejde*] "concerned with the theory/ practice interface in librarianship and devoted to the enhancement of communication between practitioners and library educators" (Kajberg, 1988, p. 148). Between 2005 and 2011, this journal was renamed *Danish Library* [*Dansk Biblioteksforskning*] and in 2012, it was taken over by the RSLIS and renamed the *Nordic Journal of Information*

[35]A history of NORDDOK and NORDINFO can be found here http://southernlibrarianship.icaap.org/content/v02n01/hannesdottir_s01.html.

Studies and Cultural Communication [*Nordisk Tidsskrift for Informationsvidenskab og Kulturformidling* (*NTIK*)[36]]. It publishes peer-reviewed articles in English and in Scandinavian languages and is poised to publish a special issue in the winter of 2018 to mark the 100th anniversary of library education in Denmark.

In 1998, Pertti Vakkari and Peter Ingwersen founded NordISNet with support from the Nordic Academy for Advanced Study (NorFA, presently NordForsk). In 2003, NordISNet was renamed "Nordic Research School of Library and Information Science" (NORSLIS). According to its website, it aims "to achieve synergic effects, effectiveness and efficiency in doctoral education and research by collecting the scattered scientific expertise, often in small-size research educational units, throughout the Nordic and Baltic countries".[37] NORSLIS organises successful PhD research courses and workshops once a year, consisting of intensive lecture series on theoretical approaches and research methods in LIS given by internationally known scholars. Doctoral students from Nordic and Baltic countries are also offered travel support to participate in these research courses or to take part in international conferences and summer schools.

Within the folk high school tradition, courses in librarianship are organised at the Nordic Folk high school in Kungälv since 1947 based on a close contact between library associations and libraries in the Nordic countries.[38]

Scholars from one Nordic country were often called to set up LIS training programmes or to chair LIS research centres in the neighbouring countries, in a spirit of positive emulation and competition. However, this mobility seemed to be mostly northwards and was especially visible in the heading of national education and research centres as well as that of Nordic agencies for education and research in LIS. Thus, the Swedish centre for Library Research (CBI) funded by the Swedish Council for Planning and Coordination of Research at the University of Gothenburg in 1987 was first headed by the Swedish Romulo Enmark until 1991. Its directorship then passed on to the Danish scholar Erland Munch-Petersen from the RSL in Copenhagen who was its acting professor between 1991 and 1994. In 1994, Lars Höglund, a Swedish scholar from the Inforsk group at Umeå University became the first full LIS professor in Gothenburg-Borås and director of this centre. Romulo Enmark then went on to become the general

---

[36]See also https://tidsskrift.dk/ntik/about.

[37]According to its website: Nordic Research School in Information Studies http://www2.abm.uu.se/norslis/. Accessed on 6 July 2017.

[38]http://www.danishfolkhighschools.com/.

secretary of NordInfo (Nordic Council for Scientific Information) between 1991 and 1993. NordInfo is in charge of fostering and funding joint research projects in LIS between Nordic countries.[39] After heading NordInfo, Enmark moved on to become the Director of the SSLIS in Borås between 1993 and 1997, and then pro vice-chancellor of the University of Borås from 1997 to 2000.

The Norwegian Research Council (NRC) also invited Enmark to chair the Norwegian Research Council on library research between 1995 and 2001. To obtain funding from the NRC, the three LIS institutions in Norway were required to tender joint research proposals. This forced their heads – Ragnar Andreas Audunson for Oslo University College; Ingeborg Solvberg for Trondheim Technical University and Niels W. Lund for Tromsø University, to collaborate and to hold a number of joint conferences.

Niels W. Lund, from the Aalborg branch of the RSL in Denmark went up to Norway to create and head the *Dokvit* programme in Tromsø between 1996 and 1999.

The annual meetings of heads of library schools in Nordic countries were presided over by the Head of the RSLIS in Copenhagen which at the time was the largest LIS school in the Nordic countries, followed by the head of the Library school in Oslo and Akershus University College. The head of the SSLIS also chaired Nordic library school meetings. The heads of these institutions wielded considerable influence over initiatives to develop LIS research.

This strong tradition of Nordic cooperation is also noticeable in mentoring university faculty in LIS. Peter Ingwersen from the RSLIS in Copenhagen had strong ties with Finnish scholars and in particular with Pertti Vakkari at Tampere University in Finland. With support from NorFA, he held various academic positions in Finnish universities. He was Affiliate Professor of scientometrics at Åbo Akademi University from 1996-2015 and Guest Professor of Information-seeking and Interactive Information Retrieval at Tampere University between 1999 and 2002. Between 2006 and 2010, he was LIS Professor at Oslo University College in Norway, and lastly between 2011 and 2012, he was Chair of Excellence in LIS at Carlos III University Madrid, Spain.

[39]For a history of NORDINFO, see Sigrún Klara Hannesdóttir's (2000), "NORDINFO: Research and Academic Library Cooperation across national borders", *Journal of Southern Academic and Special Librarianship*. Online at http://southernlibrarianship.icaap.org/content/v02n01/hannesdottir_s01.html. Accessed on 1 July 2017.

Birger Hjørland of the RSL in Copenhagen (Denmark) defended his doctoral dissertation at the SSLIS in Borås-Gothenburg (Sweden) in 1993, thus becoming the first LIS candidate to obtain a PhD in these three Nordic countries. He was also a Professor at the SSLIS between 2000 and 2003. Recall that the RSL could not deliver doctoral programmes until 1998 when it became the RSLIS with a full university status.

3.4.2. *The Anglophone Cousinry Is Well and Truly Alive*

The development of LIS in Nordic countries was influenced mostly by research and training traditions from anglophone countries. As Ole Harbo (2000, p. 6) acknowledged, "There is a long tradition of professional and academic contacts between the RSLIS and educational institutions in the other Nordic countries as well as in the United Kingdom and the United States". This is not surprising if we consider ancient history: the adventurous Germanic Saxon tribes and warring Vikings from Scandinavia settled in medieval England from the fifth century A.D.

Nordic LIS scholars are very active in international scientific and professional associations such as the IFLA, ASIST and the now defunct ASLIB. The first Rector of the RSL in Copenhagen, Preben Kirkegaard (1913–2002) who held this position from the creation of the school in 1956 till 1983, was very active on the national and international arena. He was president of IFLA from 1974 to 1979 (Harbo, 2000, p. 7); he was co-editor of the international journal *Libri* from 1957 to 1980; he was appointed Honorary Fellow of the Library Association in England.[40] Ole Harbo, Ragnar Andreas Audunson and Tom Wilson were very active in EUCLID.[41] The latter, a Professor from Sheffield (UK) was recruited as Professor at the Borås SSLIS. The open-access journal *Information Research* which he founded is still hosted by Borås University.

Nordic LIS scholars have created some of the renowned international conferences either alone or in collaboration with their anglophone colleagues. As Director of the RSL in Copenhagen, Ole Harbo was invited to attend the First International Research Forum on Information Science which took place in London in 1975. He was subsequently asked by Bertram C. Brookes, a British pioneer of IS to organise the next

[40]A brief biography can be found here http://www.gravsted.dk/person.php?navn=prebenkirkegaard. Accessed on 29 July 2018.

[41]European Association for Library and Information Science Research and Education. For a history, visit http://euclid-lis.eu/about/history/. Accessed 6 July 2017.

International Research Forum in Copenhagen in 1977 (Harbo, 2000, p. 8). The first COLIS conference held in Tampere in 1991 was organised jointly by the Finnish scholar Pertti Vakkari and the Anglo-American scholar Blaise Cronin to commemorate the 20th birthday of the Department of Library and Information Science in Tampere. The COLIS conferences have since gained a strong international reputation and attract LIS scholars from all over the world. In 1996, Perti Vakkari together with Tom Wilson coined the concept of "Information Seeking in Context" (ISIC) to mark the 50th anniversary of Finnish education in librarianship" (Mäkinen et al., 2016). This gave rise to a field of research in information behaviour with its own community and conference.

The first edition of a British-Nordic Conference on Library and Information Studies was held in Copenhagen in 1995. It was during this conference that Birger Hjørland and Hanne Albretchsen (1995) presented their landmark publication entitled "Information Science education from the point of view of domain analysis" which launched domain analysis as a competing theoretical paradigm for LIS. Domain analysis continues to have an international following and impact as a theoretical foundation for research in LIS.

Ragnar Andreas Audunson coined the concept of "public library in space" (Aabø S. & Audunson R., 2012[42]) while Roland Hjerppe in Linkoping (Sweden) tried to promote the concept of "generalised document" which is reminiscent of the semiotic conception of document by Suzanne Briet.

At the institutional level, strong cooperation existed between Nordic LIS scholars and the Rutgers school of Information in the US and with the Information Studies department in Aberystwyth[43] in Wales (UK). Peter Ingwersen had a long tradition of research collaboration with Anglo-American IS scholars working on IR and on user-centred information behaviour amongst whom Keith van Rijsbergen at Glasgow University (Scotland), Peter Willett at Sheffield University (UK) and mostly Nick Belkin at the iSchool, Rutgers University (USA). Ingwersen has been the recipient of many important awards from learned societies in LIS worldwide. In 1993, he received the Jason Farradane award from the Institute for Information Scientists in the UK; in 1994, he was awarded the

[42]Aabø and Audunson (2012).

[43]Founded in 1964 as the College of Librarianship Wales (CLW), it became part of the University of Wales Aberystwyth in 1989 and was renamed as the Department of Information and Library Studies (DILS). More information can be found here: https://www.aber.ac.uk/en/dis/latest-news/newsarticle/title-163979-en.html. Accessed on 6 July 2017.

Distinguished Lectureship Award from ASIST for his "contributions to cognitive approach to IR"; in 2005, he won the "Derek de Solla Price Medal" for his pioneering work in webometrics; in 2015, he was the recipient of the UKeiG Tony Kent Strix Award, and in 2016, he was recognised anew by the ASIST's Award of Merit, the highest award given to scholars by this society on their lifelong contributions to the field of LIS.

Likewise, the coffee-break meeting of the English Michael Buckland and the Australian Boyd Rayward with the Danish scholar Niels W. Lund at a COLIS conference in Copenhagen in 1996 launched the Document Academy, a neo-documentalist movement with its own conference (DOCAM). This community has since gained an international following of scholars interested in document theory. Its annual meetings are held in different locations all over the world.[44]

---

[44]More details can be found here http://documentacademy.org/?annual-meetings.

Chapter 4

Emergence of LIS in Spain and Portugal under Francophone Influence

Although there is a significant amount of publications on the history of LIS in Spain, they are mostly in Spanish.[1] Hence, our account of the emergence of LIS in Spain is based mostly on our oral interviews augmented by the translation of a few publications in Spanish. For similar reasons, we only offered a cursory treatment of the emergence of LIS in Portugal. Attempting to cover the emergence of LIS in Ibero-American countries is outside the scope of the current book. López Yepes and Sabelli (2013) offered a glimpse of the richness of publications and of the strong cooperation between Ibero-American countries in LIS.

The emergence of LIS education in Spain reflected the same pattern observed in the other European countries: the earliest formal training in librarianship was offered in vocational library schools outside universities. In a second phase, these special library schools began some form of collaboration with university departments to broaden their scope and incorporate academic content into their programmes. In a third phase, the vocational programmes were absorbed into university departments which gradually redesigned LIS curricula to align them with university degrees. Sections 4.1 and 4.2 are an account of this move towards universities. The contributions of the major pioneering figures of LIS in Spain are summarised in Section 4.3, while the francophone influence on Spanish LIS is examined in Section 4.4. The effect of Scientific and Technical Information (STI) policies on LIS is discussed in Section 4.5. Section 4.6 gives a brief account of the LIS landscape in Portugal.

[1]See, for instance, the selected bibliography of Félix Sagredo Fernández and Antonia García Moreno (1998) on the History of IS in Spain where all the articles they summarised were in Spanish.

4.1. From Special Library Schools to Collaboration with Universities

The first formal education in Library Science began at the "Higher School of Diplomatics" [*Escuela Superior de Diplomática*[2] (ESD)] created in Madrid in 1856 by Royal decree. This school is not to be confused with "the School of Diplomacy" [*La Escuela Diplomátic*[3]] which bears a very similar name and was also created in Madrid in 1942 almost a century later to train Spanish diplomats for service in the Ministry of Foreign Affairs and Cooperation.

The word "diplomatica" in the ESD refers to an auxiliary discipline founded in the seventeenth century by the French Benedictine monk, Jean Mabillon, author of "*De Re Diplomatica*". In this manuscript published in 1681, Mabillon laid out rules on how to determine the authenticity of old manuscripts.[4] The ESD was a vocational school run by the Spanish National Library [*Biblioteca Nacional de España*] to train archivists, specialists of old manuscripts, librarians and museum curators.

The origins of the Spanish LIS can therefore be said to be rooted in the fields of History, Palaeography, Museology and Archival Studies. Admission to the ESD was open only to civil servants who already had degrees in other areas but wished to work in public libraries and archives.

The "Public Servant Graduate for Archives, Libraries and Museums" [*Cuerpo Facultativo de Archiveros, Bibliotecarios y Museólogos*], a professional association, was in charge of safeguarding the rights and privileges of the professionals trained by the ESD.

In 1863, a new training course called "University Chair of Bibliography and Bibliology" [*Cátedra de Bibliografía y Bibliología*] focused on librarianship was run jointly by the ESD and the Universidad Central in Madrid.[5]

---

[2]https://es.wikipedia.org/wiki/Escuela_Superior_de_Diplomática

[3]https://en.wikipedia.org/wiki/Diplomatic_School_of_Spain

[4]Source: https://en.wikipedia.org/wiki/Jean_Mabillon. Accessed 21/08/2017. Diplomatics and palaeography overlap to the extent that both were concerned with the study of old manuscripts. However, diplomatics is concerned with the content, style and thus on questions of authenticity (given that many old manuscripts were forged) while palaeography is more concerned with the physical writing of old manuscripts.

[5]The history of this university dates back to the thirteenth century (1290) when a *Studium Generale* (as universities were called in those days) was created in Alcala de Henares. After being expanded, it was renamed Complutense University in 1499 after "*Complutum*", the Latin name of Alcala de Henares, the place where it was located. In 1836, the Queen Regent Maria Christina ordered the university to be moved to Madrid and it was renamed "Universidad Central" in 1851. Its original name of "Complutense University" was restored to it in the 1970s. https://en.wikipedia.org/wiki/Complutense_University_of_Madrid.

While the ESD school focused on vocational training for archivists, museum curators and palaeographers, the Universidad Central offered broader academic courses, including a few courses in library and archives. The professors at the Universidad Central carried out research whereas the ESD instructors restricted themselves to teaching. From 1900, the administration of the ESD was transferred to the Universidad Central. Until the twentieth century, Librarianship, Palaeography and Archival Studies were considered as auxiliary academic fields and not as proper "science".

For Miralpeix and Abadal (2000, p. 45), the "School of librarians" [*Escola de Bibliotecàries* (EB)] established in Barcelona in 1915 by the autonomous government of Catalonia is the first genuine library school in Spain. The EB was placed under the direction of Eugeni d'Ors and offered a three-year training programme that produced assistant librarians, thus at a lower level than the ESD in Madrid. Its programme was inspired by the new modern teaching centres in the US and the UK focused on library services to the public. It trained librarians to work in public libraries that had begun to appear in Catalonia from 1918. Its teaching staff came from both the academia and the profession. Although its three-year course was not initially recognised as a university level course, it acquired recognition within the profession both nationally and internationally (Miralpeix & Abadal, 2000, p. 45). By the first half of the twentieth century, the EB had launched two publications: a monographic series called *Workbooks* [*Quaderns de treball*] still in publication and the journal "Library management" [*Biblioteconomia*].

The orientations of these two library schools were quite different. The ESD in Madrid was accessible only to civil servants who already held a prior degree in a given field and its course content was broad and spanned Librarianship, Archival Studies and Palaeography. The EB in Barcelona did not have disciplinary entry requirements. However, admission was open only to women until 1975 because in those days, there was a widely held belief that being a librarian was a woman's job. This situation contrasted with France at the same period where the profession was dominated by the masculine gender.

With the outbreak of the Spanish Civil war (1936–1939) followed by Franco's autocratic regime of 36 years (1939–1975), Spain was closed to the outside world. Spanish scholars could not travel nor mingle freely with other scholars. Some went into exile mostly to South America.

The ESD in Madrid was closed down but the EB was placed under the local government, the "Provincial Government of Barcelona" [*Diputació Provincial de Barcelona*] and survived. Before the civil war, there were two important higher education institutions in Spain: Universidad Central (today's Universidad Complutense) and the "Center for Historical Studies" [*Centro de Estudios Históricos*] created in 1910 to produce the intellectual elite with the material means and resources to carry out research. The

Centro de Estudios Históricos supported the development of Humanities studies especially in philology. It later became the "Spanish National Research Council" [*Consejo Superior de Investigaciones Científicas*[6] (CSIC)].

Until the last quarter of the twentieth century, the ESD in Madrid and the EB in Barcelona were the only established training programmes in LIS. The Spanish National Library in Madrid offered a minor training programme which resulted in 1968, in the "National School of Documentalists" [*Escuela Nacional de Documentalistas*]. The University of Navarre also offered a short-term programme for a few years which was only open to its own librarians. This scarcity of library schools at the time meant that many professionals who worked in public and university libraries had a first degree in another field and either trained themselves on the job or through professional associations like the *Escuela de Documentalistas* (Miralpeix & Abadal, 2000, p. 45).

Save for the few library courses offered by Universidad Central in Madrid in collaboration with the ESD, LIS education was done mainly outside the university until the last quarter of the twentieth century. However, since these courses were part of other humanities courses, they did not prepare people adequately for the profession of librarian.

At the time, documentalists in Spain suffered from a lack of recognition compared to the good reputation enjoyed by information scientists or librarians. Two reasons can explain this negative perception: the inadequacies in the training programmes of the time and the lack of legal provisions and regulations for documentalists (Caridad & Portela, 1989, p. 273).

4.2. Integration of LIS Education into Universities

In 1978, the first university degree in LIS, the "Diploma in Librarianship and documentation" [*diplomatura en Biblioteconomia Documentación*], an undergraduate three-year programme was accredited by the Ministry of higher education. This paved the way for the recognition of Librarianship and Documentation as an academic field in the Spanish higher education in 1980 (Miralpeix & Abadal, 2000, p. 45) and also led to the emergence of "Schools of Librarianship and Documentation" [*Escuelas Universitarias de Biblioteconomia y Documentación* (EUBD)].

A one-year postgraduate course called "Specialisation course in documentation" [*Curso de especialización en documentación*] was created in 1981 within the Faculty of Information Science of Complutense University (in

[6]The High Council for Scientific Research.

reality the Faculty of Journalism and Media Studies). With the official recognition of the field, the course offered by the EB in Barcelona became in 1982 the "Jordi Rubió i Balaguer University School of Library economy and documentation" [*Escola Universitària Jordi Rubió i Balaguer de Biblioteconomia i Documentació*[7]] in honour of the Catalan philologist and librarian Jordi Rubió i Balaguer (1887–1982). He was also one of the most prominent directors of the school which he headed between 1931 and 1939. The administration of this school was shared between the University of Barcelona that oversaw the academic content and the local government of Barcelona who oversaw the legal and economic aspects. In 1997, the EB gained more autonomy as a university department. In July 1999, it fully integrated the University of Barcelona and became the "Faculty of Library management and documentation" [*Facultat de Biblioteconomia i Documentació*]. Since its creation in 1915, the EB had been offering a three-year undergraduate course and with the official recognition of the field, this course was recognised as a university degree. However, remnants of the old school remained under other various names and continued to train librarians to work in public libraries and in information centres.

The official recognition of the field in 1980 by the Ministry of Education led to the emergence of other university centres in LIS following the EB's model. In 1983, the University of Granada created its LIS programme, followed by the University of Salamanca in 1987, and Murcia the following year. The University of Zaragoza joined the trend in 1989. In 1990, the two universities in Madrid, Complutense and Carlos III began simultaneously to offer similar LIS programmes. Although a one-year "Specialisation course in documentation" existed at Complutense University since 1981 within its Faculty of Information Science (in reality the Faculty of Journalism and Media Studies), it was not a fully fledged LIS programme. In 1991, the University of León established its own LIS programme, followed by the University of Extremadura in 1994. The Universities of Coruña and València both set up their LIS programmes in 1996 (Miralpeix & Abadal, 2000, p. 45).

Universities either chose to run these courses through the "University Schools of Library management and Documentation" [*Escuelas Universitarias de Biblioteconomia y Documentación* (EUBDS)] established in the 1980s or they were created in other departments or faculties of these universities (Miralpeix & Abadal, 2000, p. 46).

The 1990s appear to have been a "prodigious decade" for LIS in Spain (López Yepes, 2001). It was the period when the field came to maturity and was finally recognised as an academic field as reflected in the creation of many

[7]The school is named after the Catalan philologist and librarian Jordi Rubió i Balaguer (1887–1982).

bachelor's and doctorate degrees, in the recruitment of many professors in the field and the increase in research publications of Spanish LIS scholars. In 1992, a graduate level LIS degree [*Licenciatura en Documentació*] organised as a two-year programme was open not only to students who had completed the undergraduate degree in librarianship and documentation [*Diplomatura en Biblioteconomia Documentación*] but also to people with a first degree in other areas. According to Miralpeix and Abadal (2000, p. 47):

> The *licenciatura* was offered for the first time in 1994 in four Spanish universities. Three of them already had established the *diplomatura*: Carlos III de Madrid, Granada and Salamanca. These were also joined by the Universidad de Alcalá also located near Madrid. The Complutense University of Madrid has offered the *licenciatura* since 1996 and the universities of Extremadura and the Politècnica de València, since 1997. The following year the Universidad de Murcia joined the growing ranks. In Catalonia the situation was somewhat different. Considering the long history of the School of Barcelona and despite the huge professional demand for the LIS second degree level, it seems surprising that the programme was not inaugurated until 1998. However, it was not possible to establish the *licenciatura* until the old School had become an official centre of the Universitat de Barcelona which, due to a number of circumstances, did not take place until 1997. The following year, the *licenciatura* was implemented straight away. And very recently, during the 1999–2000 academic year, two other Catalan universities, the Universitat Autònoma de Barcelona and the Universitat Oberta de Catalunya have also introduced the *licenciatura*, the latter one uniquely offering it through distance learning.

The establishment of the two degrees covering the undergraduate and graduate levels brought the legitimity and continuity needed to establish LIS faculties and departments in universities. It enabled the last library schools, the EUBDs to become universities faculties. It also facilitated the development of doctoral programmes in the field.

However, the two oldest LIS schools, Barcelona and Madrid maintained their dominant position (Miralpeix & Abadal, 2000).

Unlike what happened in France, professional associations played a significant role in shaping LIS education programmes. Librarians and archivists working at the highest level in public administrations were also members of professional associations such as the "Public Servant Assistant for Archives, Libraries and Museums" [*Cuerpo de Facultativos de Archivos, Bibliotecas y Museos del Estado*]. The "National Association of Archivists, Librarians and Archeologists" [*Asociación Nacional de Archiveros, Bibliotecarios y Arqueólogos* (ANABA)] was created in 1949 to unite the different information professionals under one umbrella organisation.

"Documentalists" was later added to the name to make it "ANABAD". Today this association is a "Federation of Spanish Associations of Archivists, Librarians, Museologists and Documentalists" [*Féderación Española de Asociaciones de Archiveros, Bibliotecarios, Museólogos y Documentalistas*]. In 1976, the "Spanish Society of Documentation and Scientific Information" [*Sociedad Española de Documentación e Información Científica* (SEDIC)] was created in Madrid to cater for professionals working in information centres. Its Catalonian counterpart, the "Association of Librarians and Documentalists of Catalonia" [*Societat Catalana de Documentació i Informació* (SOCADI)] was founded in 1984. When it merged with the "Official Collegiate of Librarians-Documentalists of Catalonia" [*Collegi Oficial de Bibliotecaris-Documentalistes de Catalunya*] in 1998, it became "the library association with the highest membership in Spain with some 1500 members" in the 2000s (Miralpeix & Abadal, 2000, p. 45). These associations were invited by the Ministry of Higher Education to make recommendations on the reform of LIS education. In particular, the SEDIC and the ANABAD played a major role in re-structuring the documentation curriculum as a three-year undergraduate programme followed by a two-year bachelor's degree such that the whole curriculum took five years to complete and corresponded to the current master's degree.

As LIS education made its way into universities, it was faced with a new challenge in the form of academic territory staking or academic guerrilla warfare in which institutions of higher education excel. As it happened, the name "information science" had already been appropriated by journalism and media studies and could therefore not be used by LIS scholars to name their courses, departments, schools and indeed the field. The Faculty of Information Sciences[8] created in 1971 at the Complutense University in Madrid was in reality a Faculty of journalism and audiovisual communication. Since journalism was not considered a "science", calling itself "information science" (IS) conferred upon it a "scientific" aura and made it more acceptable as an academic field.

LIS scholars were thus left with the option of naming their education institutions either as "documentation or library management". This explains why the University of Barcelona (UB) has a "Faculty of Library management and Documentation" [*Facultat de Biblioteconomia i Documentació*]; the Complutense University has a "Faculty of Documentation sciences" [*Facultad de Ciencias de la Documentación*[9]] while the LIS unit at Carlos III

[8]"Facultad de Ciencias de la Información". See their website for a history of this faculty (in Spanish) http://ccinformacion.ucm.es/historia.

[9]The 's' in sciences is because implicitly they train archivists and librarians as well, so it is actually a Faculty of Documentation, Library and Archival studies. http://documentacion.ucm.es/.

University also in Madrid is called "Department of Library management and Documentation" [*Departamento de Biblioteconomía y Documentación*[10]]. These faculties offer a full LIS programme going from undergraduate to doctorate level.

Similar to the trend observed in other countries like Portugal, the UK and the US, in the last decade, some LIS departments in Spain are being merged with Communication departments for economic reasons.

Figure 4.1 here below shows the location of the earliest training programmes and how they were gradually absorbed into universities until the official recognition of the field in the 1980s.

4.3. Spanish Pioneers in LIS and the weight of the Complutense School in Madrid

According to López Yepes (2001), the Spanish documentation school was initiated by the Spanish philosopher Ortega y Gasset. Although Ortega did not use the words "documentation" or "document" in an important speech he gave in 1935 entitled "Mission of the Librarian" [*Misión del Bibliotecario*], the idea was present.

Other pioneers who played a prominent role in the emergence of the LIS discipline included Agustín Millares, Javier Lasso-de-la-Vega Jiménez-Placer, José Maria Desantes Guanter, José López Yepes, Félix Sagredo Fernando, José-Ramón Pérez Alvarez-Ossorio and Emilia Currás. The majority of these scholars trained and worked at the Complutense University of Madrid.

Agustín Millares Carlo[11] (1893–1980) was a historian, a palaeographer and an archivist who taught at the ESD in Madrid before the civil war in the 1930s. Millares can be said to have given an impetus to the development of archival studies and bibliography in Spain. He made the link between the ESD, the Centro de Estudios históricos and the Universidad Central. At the outbreak of the Spanish Civil war, he was forced to go into exile to México where he taught at the Universidad Nacional Autónoma de México (UNAM) and then to Venezuela at the Universidad del Zulia, Maracaibo, during Franco's regime (Moreiro, 1989).

Javier Lasso-de-la-Vega Jimenez-Placer[12] (1892–1990) taught library studies, economics and library management at the vocational school

[10]The webpage is here: http://www.bib.uc3m.es.

[11]A short biography in Spanish can be found here: https://www.biografiasyvidas.com/biografia/m/millares.htm.

[12]A very short biography in Spanish can be found here: https://es.wikipedia.org/wiki/Javier_Lasso_de_la_Vega_Jiménez-Placer.

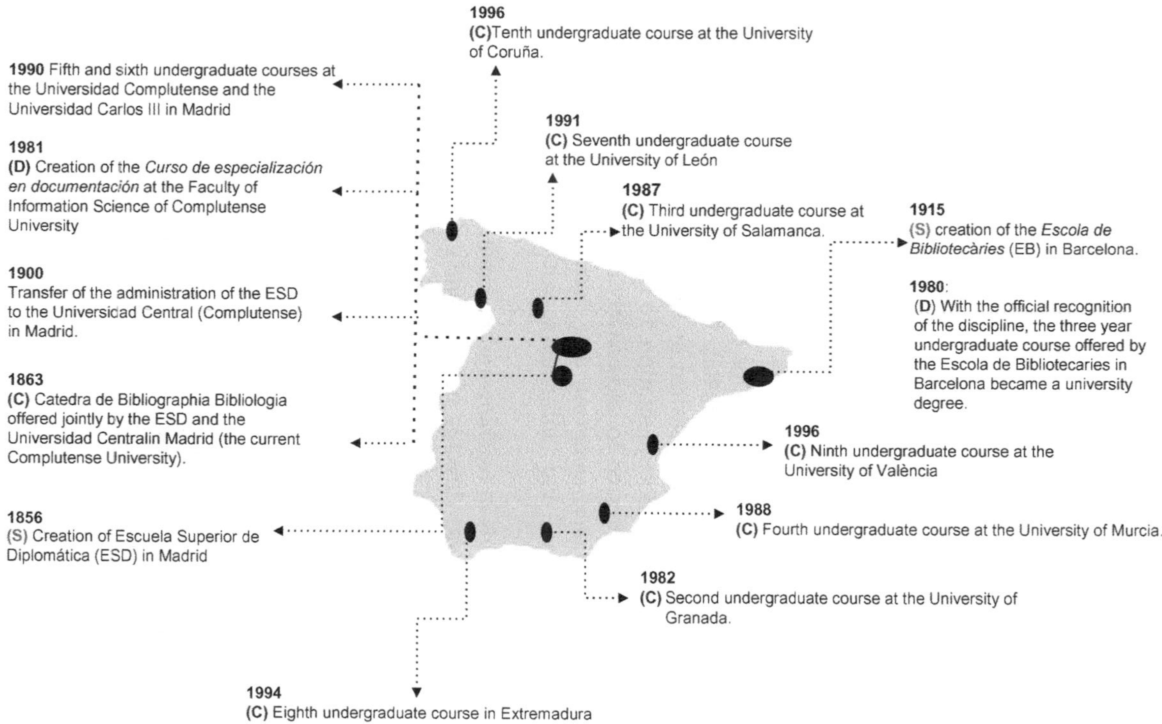

Figure 4.1: LIS education programmes in Spain and the importance of the Madrid and Barcelona schools.
Key: C = Course; D = Diploma/Certificate; S = School; I = Infrastructure.

"*Escuela Nacional de Documentalistas*" in Madrid and later at the theology faculty of the University of Salamanca. He is considered the founding father of Spanish documentation school (Lopèz-Yepes, 2001). The term "scientific documentation" appeared in 1947 in one of his articles entitled "Documentation in medicine. International medical bibliography 1945-51" [*Documentaciòn en medicina. Bibliografía médica internacional 1945-51*]. Jiménez-Placer also wrote a manual in 1947 on "How to do a doctoral thesis, or technical manual of scientific and bibliographic documentation" [*Còmo se hace una tesis doctoral, o manual de técnica de la documentaciòn científica y bibliográfica*]. In his 1954 publication, "Librarian and documentalist. A friction and a problem" [*Bibliotecario y documentalista. Una fricciòn y un problema*], Jiménez-Placer used the term "documentation" to name the discipline. He retraced the genesis of the concept of documentation by comparing its different definitions in the literature. In 1969, he wrote the first documentation manual in Spain entitled *Manual of Documentation: Techniques for researching and writing scientific and engineering works* [*Manual de Documentaciòn. Las técnicas para la investigaciòn y redacciòn de los trabajos científicos y de ingeniería*] (Abadal, 1992, 260). In a later edition of this Manual of Documentation published in 1977, an entire chapter was dedicated anew to definitions of documentation, to the name of the discipline and its professionals that he referred to successively as "documentist" [*documentista*], "documentalist" [*documentalista*] and "documenter" [*documentador*], "information technicians" and "informative scientists" (Abadal, 1992, pp. 160–161).

Aside from his contributions in teaching and research, Jimenez-Placer was also an administrator. He was director of the library at the Complutense University where he worked on coordinating and normalising university library catalogues using the Universal Decimal Classification (UDC). He held various positions in professional and scientific LIS associations: he was president of the "Spanish Association of Friends of Libraries" [*Asociación Española de Amigos de las Bibliotecas*], he was active in the "Public Servant Assistant for Archivists, Librarians and Archeologists" [*Cuerpo Facultativo de Archiveros, Bibliotecarios y Arqueólogos*] as well as in the "Spanish National Research Council" [*Consejo Superior de Investigaciones Científicas*]. He worked with international organisations such as the UNESCO and the Paris-based FID on the design of curricula in documentation and library studies.

The third pioneer, José María Desantes Guanter[13] (1924–2004) can be said to be the mentor of the Complutense LIS school. After a PhD in Law and

[13]A short biography in Spanish can be found here: https://es.wikipedia.org/wiki/José_Maria_Desantes_Guanter.

Information Science at this university, he went on to become a professor there. He brought documentation into the IS curricula and worked on digital media and audiovisual documentation for the press. As an expert in the field of law and ethics of information, he published extensively on the legal aspects of documentation. In particular, his "Theory and legal regime of documentation" [*Teoría y régimen jurídico de la documentación*] published in 1987 is considered a classics on the topic. Desantes Guanter produced a string of doctorate students who are currently LIS professors in this university.

The fourth pioneer Félix Sagredo Fernández is a versatile scholar who was also a chemist and a priest. He obtained his first university position teaching history. A contemporary of José López Yepes, Sagredo Fernández became a professor of Information Science at Complutense University. He headed the huge LIS programme in that university for several years. He co-authored a book with José-María Izquierdo Arroyo entitled *Logico-linguistic conception of documentation* [*Concepción Lógico linguística de la Documentación*] in 1983. Sagredo Fernández and Antonia García Moreno's 1998 article "History of information science in Spain: A selected bibliography" offered a glimpse to non-Hispanic speakers of the amount of publications in Spanish on different aspects of the history of IS between 1952 and 1995.

José-Ramón Pérez Álvarez-Ossorio's contribution to the development of LIS in Spain lay mostly in the professional sector. In 1956, he joined the "Information and Documentation Center" [*Centro de Información y Documentación* (CID)] and became its director in 1968. The CID became the "Institute of Information and Documentation in Science and Technology" *Instituto de* [*Información y Documentación en Ciencia y Tecnología*] in 1975.

The only woman amongst the seven pioneers, Emilia Currás Puente (1927-) was a chemist who trained at Complutense University and in Germany. She became professor at the Universidad Autónoma of Madrid. Her areas of research were in philosophy and history of science, documentation and knowledge organisation. She contributed to the philosophical basis of information science and knowledge organisation.[14] Currás was also very active in professional LIS associations both at home and abroad. She founded the Spanish professional association SEDIC in 1976. She was also a one-time president of the Spanish chapter of the International Society for Knowledge Organization as well as an active member of FID.

A disciple of José María Desantes Guanter, José López Yepes holds a PhD in philosophy and in Spanish literature from the Complutense University where he also later became a professor in the Faculty of Information Science. Both men were instrumental in setting up the first documentation course at this university. López Yepes's 1978 book entitled

[14]A short biography can be found: https://es.wikipedia.org/wiki/Emilia_Currás.

Teoria de la documentación is one of the first Spanish texts where a systematic and unified approach to documentation is attempted. He was the founder and editor-in-chief of the following journals: "*Documentation of Information Sciences*" [*Documentación de las Ciencias de la Información*], "*General Journal of Information and Documentation*" [*Revista General de Información y Documentación*] and *Journal of Spanish Research on Information Science*.[15]

The other LIS scholars who make up the Madrid group are Alfonso López Yepes[16] (not to be confused with José López Yepes), Mercedes Caridad Sebastían, Antonio García Gutiérrez, Félix del Valle and Rosa San Segundo. The last four were doctorate students of José López Yepes[17] who many consider currently as one of the leading figures in the Madrid LIS school. Mercedes Caridad Sebastían was also a one-time president of SEDIC, the Spanish association of Documentation and Information Science.

In the years following the recognition of the field in 1980, the research landscape of LIS was fragmented, embryonic and lacking in visibility. Since then, Spanish LIS scholars have become quite visible on the international research arena particularly in Latin American. Spanish pioneers were also active in international LIS associations such as FID, IFLA and UNESCO. In particular, José López Yepes, Emilia Currás, and José-Ramón Pérez-Álvarez-Ossorio were all active in FID which was the link with other European countries. One of the annual meetings of the FID was organised by the "Institute of Science and Technology of the Spanish National Research Council" [*Instituto de Ciencia y Tecnología del Consejo Superior de Investigaciones Científicas* (ICYT-CSIC)] in October 1992 in Madrid.

4.4. The Francophone Influence and the Shadow of Paul Otlet

There is an overwhelming consensus that the emergence of LIS in Spain was influenced by the francophone documentation tradition:

> Spain, during the period that information science was developing in the Anglo-American world, saw documentation emerge as a discipline, including research in the nature of document (Ros García, 2000). Spanish thinking has explicitly based itself on the theoretical works

[15]A bibliography in Spanish can be found here: https://elblogdejoselopezyepes.wordpress.com/perfiles-de-investigadores/dr-jose-lopez-yepes/.

[16]A short biography can be found here: http://iibi.unam.mx/micrositio/CIADS/ciads_mesa6-4_lopez.pdf.

[17]A list of students whose PhD he supervised can be found here http://pendientedemigracion.ucm.es/info/multidoc/profesores/j_lopez.htm.

> by Otlet and Briet (Martínez-Comeche, 2000, pp. 5–10). The relationship between information, communication, and documentation is a matter of discussion. (Lund, 2009, p. 406)

The fact that French was the first foreign language taught in Spain at the time helps explain this. López Yepes (2001) observed that:

> in their search for the origins of our discipline, the Spanish academia has taken a great interest in the study of the figure of Paul Otlet and his works, thus producing the highest number of contributions worldwide.

We do not have information to substantiate the last part of his claim but Otlet and Briet's influence are undeniable in the choice of documentation to name LIS institutions in Spain. Similar to the emergence of documentation in France, this new subfield was considered "a new area of activity that expanded the role of the library" and its arrival was embraced by Spanish scholars and librarians (Yepes, 2001, p. 262). Rodríguez-Bravo, Simões, and Martínez-Ávila (2018) also observed that "the choice of the term "documentation" is greatly indebted to the French tradition".

From the 1960s, conferences, panels and congresses organised in the field began to include a documentation session. In 1961, a documentation section appeared in the bulletin of the "General Directorate of Archives and Libraries" [*Dirección General de Archivos y Bibliotecas*] and in 1969, a Bulletin of Documentation was created by the "Funds for the Economic and Social Research of the Confederate Savings Bank" [*Fondo para la Investigación Económica y Social de las Cajas de Ahorros Confederadas*]. Documentation continued to gain ground as the core concept of LIS in the 1970s both through its theoretical exploration and in the multiplication of LIS education programmes and Faculties in universities. In 1972, an article by Nuria Amat entitled "About Documentation and documentary treatment" [*Sobre la Documentación y tratamiento documental*] shone a light on two crucial elements that strengthened the documentation theory: the re-discovery of Paul Otlet's work and the introduction into Spain of the ideas of the North American IS pioneer Harold Borko whose 1968 definition of information science became a classic in the field.

José López Yepes (1978)'s "Theory of documentation" [*Teoria de la documentación*] analysed the influence of Paul Otlet on the Spanish documentation movement. The University of Zaragoza organised the "Commemorative days of the fiftieth anniversary of Paul Otlet (1868–1944)" [*Jornadas conmemorativas del Cincuentenario de Paul Otlet (1868–1944)*] in 1995. The influence of the documentation tradition reached its peak with the edition of "Treatise on documentation" [*Tratado de Documentación*] by María Dolores Ayuso in 1996. Félix Sagredo Fernández worked with Otlet's foundation *The Mundaneum* in Brussels and visited it many times.

Jean Meyriat, one of the founding figures of information-documentation in the "Information and Communication Sciences" in France had ties with some of the Spanish LIS scholars. In particular, Mercedes Caridad Sebastían worked with Meyriat within the European Association for Library and Information Education and Research (EUCLID).[18] Meyriat lived and worked in Madrid for some time during the Second World War.[19]

4.5. Policies for Scientific and Technical Information Infrastructures in Spain

The 1970s and 1980s were important decades for IS worldwide not only because LIS was being acknowledged as an academic field in many countries but also because of the Cold War era. Many western countries considered the development of Scientific and Technical Information (STI) infrastructures a matter of national sovereignty. Just as successive French governments had implemented various STI policies in the 1960s–1970s[20] and Germany had embarked on an "Information & Documentation" (I&D) programme in the 1970s (Hauk & Stock, 2012), Spain also embarked on a national programme of information policy in order to catch up with more advanced European countries.

In 1982, the National Plan of Action for 1984–1986 on Scientific and Technical Documentation and Information (IDOC plan) was set up by the Sub-directorate General for Documentation and Scientific Information, created within the State Secretariate of Universities (Caridad Sebastián, Méndez Rodríguez, & Rodríguez Mateos, 2001, 51). The IDOC plan was the documentation and STI counterpart of a 1986 science law named "Promotion and Coordination of Scientific and Technical Research" aimed at promoting, regulating and coordinating scientific research in Spain via several dedicated agencies and bodies such as the "Interministerial Commission of Science and Technology (CICYT), the Advisory Council for Science and Technology (CACYT), the General Council of Science and Technology and the Center for Technological and Industrial Development (CDTI)" (Caridad Sebastían

[18]See its website: http://euclid-lis.eu/.

[19]According to this biography, he taught in a high school in Madrid between 1943 and 1946. https://www.reseau-canope.fr/savoirscdi/societe-de-linformation/le-monde-du-livre-et-de-la-presse/histoire-du-livre-et-de-la-documentation/biographies/jean-meyriat-1921-2010-un-batisseur-de-linformation-documentation.html.

[20]The grand plan for computation ("plan calcul") recommended in 1964 by the Gaullist government was followed by a grand plan for telecommunications in 1974 championed by the centrist government of Valéry Giscard d'Estaing, and then by a policy for democratisation of access to information in the 1980s by the socialist government of François Mitterand.

et al., 2001). Unfortunately, like the German I&D programme, many of the recommendations made by the IDOC plan remained *lettre morte*.

4.6. Emergence of LIS in Portugal

Our brief foray into the Portuguese history of LIS relied on a synthesis of the publications of Fernanda Ribiero, a Professor in the department of Information and Communication, University of Porto and on a biographical sketch of Portuguese pioneers which she graciously prepared for us in English. A more in-depth study of the emergence of LIS in Portugal is therefore still needed.

4.6.1. From Vocational Library Schools to the University

Formal education in librarianship and archives began in Portugal in the nineteenth century and was done largely outside universities. The "Course for Librarians and Archivists" [*Curso de Bibliotecário-Arquivista*], a professional diploma for training archivists and librarians was set up in 1887. It was administered jointly by the "Higher Course of Literatures" [*Curso Superior de Letras*] founded in Lisbon in 1859 and by the National Library and National Archives in Torre de Tombo (Ribeiro, 2008). With the creation of the University of Lisbon in 1911, the *Curso Superio de Letras* was transferred to the Faculty of Arts and Humanities although technical courses continued to be managed by the National Library and National Archives. The *Curso de Bibliotecário-Arquivista* was redesigned as a three-year general course open to anyone with a first degree but unlike other university undergraduate courses, it did not lead to a bachelor's degree. Due to low student attendance, a decree promulgated in 1931 saw this course being taken out of the university and being reverted to a two-year vocational postgraduate course administered by the Directorate of Libraries and Archives [*Inspecção das Bibliotecas e Arquivos*]. Admission was open to students from the Faculty of Arts and Humanities, the aim being to complement their initial academic and theoretical education with a technical and practical course in librarianship and archival studies. Trainees were destined to become "senior officers in libraries and archives". These early university training programmes in LIS were broad in their design so as to fit both the education needs of librarians and archivists (Ribeiro, 2008).

In 1935, a two-year course, the "Higher Course for Librarians and Archivists" [*Curso Superior de Bibliotecário-Arquivista*] with a dominant historical component was established at the University of Coimbra to train librarians and archivists to work in state libraries and archives. In 1969, a new technical training programme was created outside the university and placed under the supervision of the "General Directorate for Higher Education and Fine Arts" [*Direcção-Geral do Ensino Superior e das Belas-Artes*] in order to respond to the need for librarians and archivists. The University of Coimbra's programme was more historical and custodial in its orientation, thus reflecting a European perspective whereas the technical programme of the Directorate was inspired by the American model where information science and information technologies were being pushed to the forefront (Ribeiro, 2008). This new technical programme was however closed down in 1974 following the Carnation Revolution of 25th April 1974 which deposed the authoritarian regime of the country. As in the other European countries, professional associations such as the IFLA, FID, International Conservation Association (ICA) and UNESCO played a key role in efforts to harmonise LIS education worldwide and to identify core course contents in the early 1980s (Ribeiro, 2007).

In 1982, a second two-year programme "Specialisation course in Documentation Sciences" [*Curso de Especialização em Ciências Documentais* (CECD)] was established within the Faculty of Arts and Humanities at the University of Coimbra to replace the old *Curso de Bibliotecário-Arquivista.* The CECD also marked an end to the broad educational model which had lasted almost a century by separating the training for librarians and documentalists from that of archivists. For more than half a century, the University of Coimbra was the only place where a LIS training programme could be obtained in a Portuguese university.

The CECD became the model of LIS education replicated in several Portuguese universities (Lisbon, Porto). At the time of publication of Ribeiro's 2008 article, 17 universities and polytechnic schools were offering CECD courses. There was a rapid growth in the number of trainees graduating with this diploma from 50 graduates in 1983 to around 640 in 2006. Like the French educational landscape at the same time, the CECD remained a technical course with little or no theoretical and historical background. It was open to students coming from different disciplines who were thus beginners in LIS. Also, similar to France at the same time, it was difficult to receive an entire university curriculum in LIS, from undergraduate to doctoral level (Ribeiro, 2008). This hampered the academic and scientific development of the field in Portugal.

From the 1990s, efforts were made to expand LIS education in Portugal and to incorporate more academic content into it. A number of Portuguese professors in LIS were trained in Spain and in the UK due to the tradition of cooperation between Portugal and the UK in the education, trade, industry and defence sectors (Ribeiro, 2008). This enabled the setting up of joint diplomas between the two countries. For instance, in 1991, the National Laboratory of Engineering and Technology [*Laboratório Nacional de Engenharia e Tecnologia Industrial*] in Lisbon partnered with the University of Sheffield to deliver the first Master of Science (MSc) degree in

Information Management. It is worth noticing that this master's course was not a traditional LIS programme but was aimed at training highly qualified "information intermediaries" (infomediaries) needed by the country to develop its information sector (in both public and private industries and services). Thus, the first MSc in LIS had a very utilitarian and pragmatic bent as these lines written by its Portuguese and UK coordinators show:

> The initiative is seen not only as an educational innovation, but as an attempt to introduce to a wider audience than hitherto-and particularly the business sector the concepts of modern information management, not only those relating to the techniques and technology, but also the wider issues of access to external information resources of significance to business. (Correia and Wilson, 1992, p. 81)

This MSc programme also set the trend for the creation of other master's degrees in "information-documentation", "information management and libraries" that tried to combine a broad theoretical knowledge base with professional technical skills. The first PhD programme was created at the Faculty of Arts and Humanities of the University of Porto in 1989, followed by Coimbra a few years later.

It was not until 2001 that a Bachelor in Information Science [*Licenciatura en Ciência da Informação*] was created at the University of Porto to replace the former CECD course. This BA programme was designed in partnership between two Faculties of the University of Porto: the Faculty of Humanities and Social Sciences and the Faculty of Engineering. It aimed to confer a more theoretical grounding and a broader approach to LIS education that avoids the artificial separation along professional lines (archivists *vs* librarians *vs* documentalists) while incorporating the newer information technologies. The model of this BA has been replicated in other Portuguese universities to ensure an education programme that combines technical skills in librarianship and archival studies with epistemological, historical and theoretical contents. Figure 4.2 hereafter shows a map of the various LIS programmes created in Portugal.

4.6.2. *Administrators of Libraries and Archives in the Early Twentieth Century*

Three figures played an important role in the development of libraries and archives in the early twentieth-century Portugal: Júlio Dantas, Raul Proença and António Ferrão. Júlio Dantas[21] (1876–1962) was a renowned writer and diplomat who became "Inspector of Libraries and Archives" [*Inspeção das Bibliotecas*

[21]A longer biography in Portuguese can be found on his Wikipedia page: https://pt.wikipedia.org/wiki/Juúlio_Dantas. Accessed 30th July 2017.

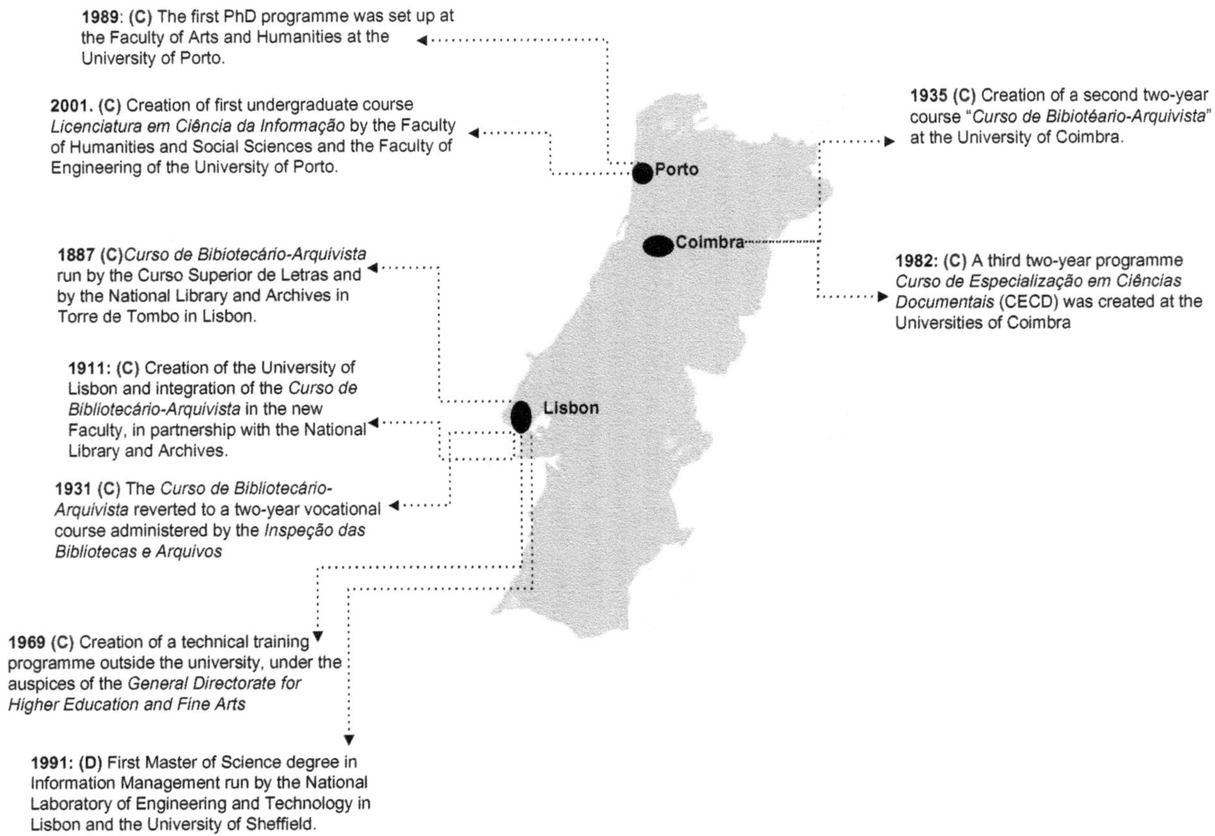

Figure 4.2: The Development of LIS Education in Portugal. *Key*: C = Course; D = Diploma/Certificate; S = School; I = Infrastructure.

e Arquivos] in 1912. He remained in this office for thirty-four years (until 1946). Although, he is better known for his literary activities, he contributed in designing political and strategic orientations for libraries and archives. While his work in the LIS sector was mainly legislative, his name cannot be dissociated from the development of public library and archival services in Portugal.

The second figure, Raul Proença[22] (1884–1941) was also a writer, a journalist and an intellectual. He became the head of Technical services of the National Library and also secretary of the journal "Annals of Libraries and Archives" [*Anais das Bibliotecas e Arquivos*]. Proença made decisive contributions to LIS in three areas: research, management and technical. On the research side, he contributed in writing a proposal for the reform of the National Library. During 1919, he worked on the elaboration of the *Cataloging Rules* which required up-to-date technical knowledge on the topic as well as knowledge of the cataloguing rules in use in other parts of the world. During the "Congress of Librarians and Bibliophiles" [*Congrès des Bibliothécaires et Bibliophiles*] that took place in Paris in April of 1923 and which was attended by more than three hundred participants from Europe and the United States, Raul Proença and Jaime Cortesão presented their work on cataloguing rules at the Portuguese National Library. This elicited the unequivocal praise of Paul Otlet, the then director of the *Palais Mondial* in Brussels and secretary of the International Institute of Bibliography (IIB). The report of their presentation in the Congress was published in the *Anais*[23] journal.

The third figure, António Ferrão[24] (1884–1961) was a historian, a diplomat and a professor at the Faculty of Literatures of the University of Lisbon. He held various public functions in the Ministry of Public Instruction[25] during the first Portuguese Republic. He was a member of the Commission on Intellectual Property and was in charge of reviewing and updating Portuguese legislation on literary, scientific and artistic property. As a *connoisseur* of the trends in vogue in his time and a thinker focused on action, Ferrão sought to translate into practical applications the ideas and analysis he developed on the many subjects he touched upon. After the creation of the Ministry of Public Instruction in 1913, Ferrão held the functions of the Head of Department and Head of Service of Artistic Instruction and University

---

[22]A Wikipedia stub in English is available here: https://en.wikipedia.org/wiki/Raul_Proença. Accessed 30th July 2017.

[23]See: CORTESÃO, Jaime; PROENÇA, Raul – O Congresso de Paris. *Anais das Bibliotecas e Arquivos*. Lisboa. 2.ª série. 4:13/14 (Jan.-Jun. 1923) 80-113.

[24]A short biography in Portuguese is available here: https://pt.wikipedia.org/wiki/António_de_Abranches_Ferrão. Accessed 30th July 2017.

[25]This was the name given to what is known today as the Ministry of Education when it was created at the beginning of the twentieth century (1913).

Instruction. Between 1913 and 1918, he actively sought to implement a series of proposals for the development of the library and archival sector.

On February 8, 1929, following a change in the ministerial organic structure, he was appointed Sub-Inspector of the Erudite Libraries and Archives, a position which enabled him to make significant contributions to the field. He instigated various changes in the legislation and contributed in publishing technical procedures for public archives and libraries as well as guidelines on how to establish different types of libraries in the country.

Ferrão also studied the organisation of archives, libraries and fine arts institutions abroad and did an inventory of documents pertaining to the history of Portugal in foreign libraries and archives. This led him to visit libraries and archival institutions in several European countries including Spain, France, Belgium, Switzerland, Italy, Germany, Belgium and the Netherlands. Ferrão was also actively involved in the education of librarians and archivists. He taught in the *Curso de Bibiotecário-Arquivista* in Lisbon. His period of greater fame came when he became the Inspector General of Libraries and Archives in 1946 following the retirement of Júlio Dantas. He remained in this position until his retirement in 1954.

The important positions held by Júlio Dantas and António Ferrão in the Ministry of Education under which libraries and archives were governed influenced the production of legislation that regulated the *Curso de Bibliotecário-Arquivista*. Moreover, António Ferrão taught in the *Curso*.

4.6.3. The Battle for Survival and for Academic Recognition

Unlike the other European countries we studied, LIS has not yet acquired academic status as an autonomous discipline in Portugal. Although LIS programmes are offered at both bachelor and masters' levels, they are integrated into other departments like History, Arts and Humanities, Communication and Media studies or, more recently, into Management. The first Bachelor in Information Science created in 2001 was lodged in the Department of Cultural Heritage alongside other programmes in Archaeology, History of Art and Museology at the University of Porto. This course was integrated into the Department of Journalism and Communication Sciences in 2007.

Until 2001, LIS courses could not use the name "information science" in their titles and until 2006, "information science" was not listed as an academic field in the National Foundation of Science and Technology directory. Since the discipline has not officially been recognised, applications for LIS research projects or scholarships are submitted to scholars from Communication Sciences or from History and thus were consequently not evaluated positively. After much lobbying from the University of Porto, in 2006, the word "information" was finally added to "communication sciences" to obtain the field of "Communication and Information Sciences".

In 2010, the University of Coimbra created four departments within the Faculty of Arts and Humanities one of which was named "Department of Philosophy, Communication and Information".

The University of Porto followed suit in 2016 by changing the name of its department of Communication Science to "Department of Communication and Information Sciences".

The situation of the discipline in Portugal bears some similarity to that of France *albeit* with the notable difference that in France, the merger with communication science happened at the birth of the "Information and Communication Sciences" (ICS) interdiscipline [*Sciences de l'Information et de la Communication*] in 1972/5. In Portugal, "IS" was added to an already existing field of Communication Sciences several decades after the first training programmes in the field emerged.

Like Spain, Portuguese LIS is very much under francophone influence. A recent study by Rodríguez-Bravo *et al.* (2018) investigated Paul Otlet's influence on Spanish and Portuguese speaking authors. A citation count showed that "more than one third (37.5%) of all citations to Otlet's work came from Portuguese and Spanish speaking authors". Portuguese authors cited Otlet twice more (26.03%) than Spanish authors (11.47%). Their study further showed that the top eight authors with the highest number of citations to Paul Otlet's works were young Brazilian LIS authors. According to the authors, the re-discovery of Otlet's works is an effort to recover "the French tradition of Library Science and Documentation that was lost due to ideological reasons, especially in Brazil" [...] "the influence of French scholars was ideologically interrupted in Brazil by an American school in the mid-20th century" (Rodríguez-Bravo et al., 2018).

While this short account of the Portuguese LIS landscape is incomplete, it offers a glimpse into the emergence of LIS in Portugal. Also, our cross-national approach highlighted the influence of the francophone documentation tradition on Spanish and Portuguese LIS as opposed to the other countries studied (Croatia, Denmark, Sweden, Norway) where a strong anglophone influence is perceptible.

4.7. Network of Links between European Pioneers of LIS

While Paul Otlet and Suzanne Briet remain central influences all over the world, the predominance of the francophone tradition on the Hispanic LIS landscape is symmetrical with that of the anglophone influence on the Nordic and Yugoslavian LIS landscapes. Figure 4.3 is an attempt to map out the network of external influences amongst European pioneers in the seven countries studied and their links with LIS scholars outside of Europe (UK, US and Australia mainly).

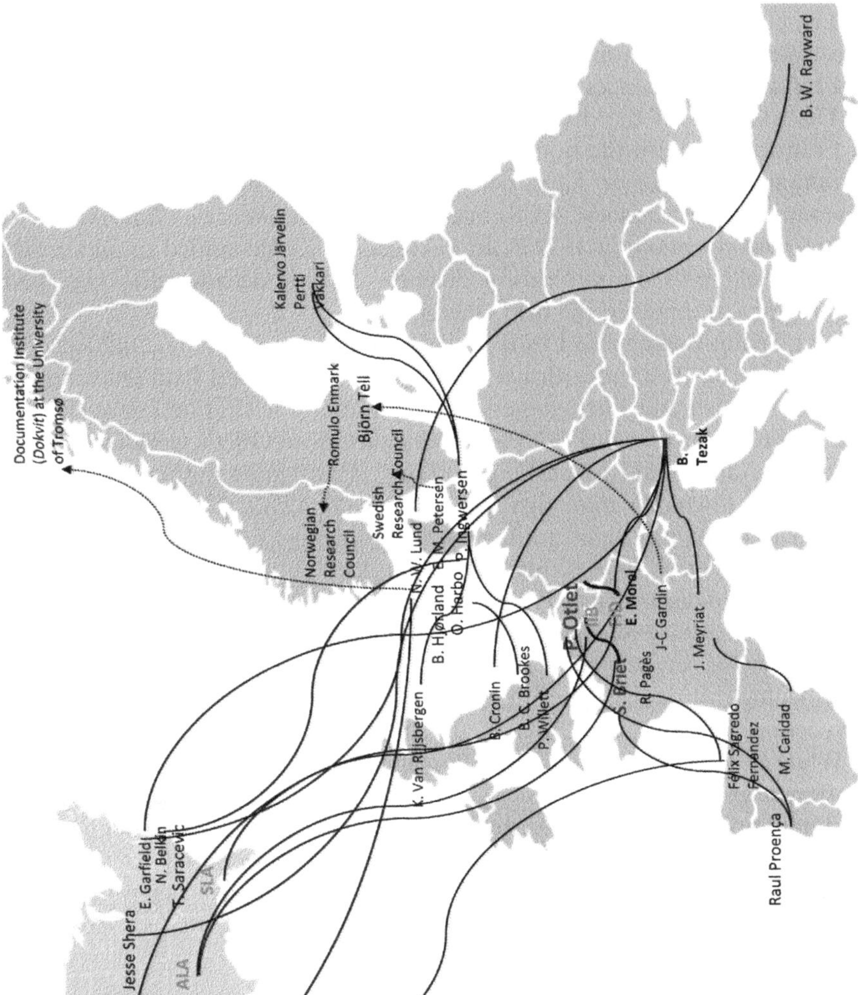

Figure 4.3: The Network of External Influences between European Pioneers, Scholars and Other Pioneers of LIS Worldwide.

The map is not meant to be an exhaustive picture of every pioneer's links, rather it shows the most significant links we uncovered in the course of our study, mainly through oral communications with European LIS scholars (interviews and email exchanges). This figure shows the importance of professional associations such as IIB, FID, ALA, and the SLA in structuring the LIS landscape of many European countries in the first half of the twentieth century.

The two tables hereafter present these links in a more linear manner. Table 4.1 shows the links between scholars in a two-column table. Each row indicates two scholars who knew each other. The first column gives the names of the European pioneers from the seven countries we studied and in the second column 2, the names of other scholars (European or not) with whom they collaborated or were acquainted.

Table 4.1: Links between European LIS Scholars and other European and Non-European Connections.

Scholar 1	Scholar 2
Eugène Morel (France)	Paul Otlet (Belgium)
Eugène Morel (France)	Henri La Fontaine (Belgium)
Eugène Morel (France)	Suzanne Briet (France)
Suzanne Briet (France)	Paul Otlet (Belgium)
Raul Proença (Portugal)	Paul Otlet (Belgium)
Félix Sagredo Fernández (Spain)	Paul Otlet (Belgium)
Félix Sagredo Fernández (Spain)	Henri La Fontaine (Belgium)
Félix Sagredo Fernández (Spain)	Michael K. Buckland (USA)
Jean-Claude Gardin (France)	Björn Tell (Sweden)
Božo Težak (ex-Yugoslavia)	Eugene Garfield (USA)
Božo Težak (ex-Yugoslavia)	Robert Hayes (USA
Božo Težak (ex-Yugoslavia)	Jesse Shera (USA)
Božo Težak (ex-Yugoslavia)	Blaise Cronin (UK & USA)
Jean Meyriat (France)	Božo Težak (ex-Yugoslavia)
Jean Meyriat (France)	Mercedes Caridad Sebastían
Ole Harbo (Denmark)	Bertram C. Brookes (UK)
Niels W. Lund (Denmark)	Boyd W. Rayward (Australia)
Niels W. Lund (Denmark)	Michael K. Buckland (USA)
Peter Ingwersen (Denmark)	Pertti Vakkari (Finland)
Peter Ingwersen (Denmark)	Kalervo Järvelin (Finland)
Peter Ingwersen (Denmark)	Tefko Saracevic (USA)
Peter Ingwersen (Denmark)	Nick Belkin (USA)
Peter Ingwersen (Denmark)	Keith van Rijsbergen (Scotland)
Peter Ingwersen (Denmark)	Peter Willett (UK)

Table 4.2: Linearises the information concerning the mobility between Nordic scholars from Denmark, Sweden and Norway.

Scholar 1 from Country 1	Appointed/Worked in Country 2
Erland M. Petersen (Denmark)	1991–1994: Head of Swedish Research Council (Sweden)
Romulo Enmark (Sweden)	1991–1993: General secretary of NORDINFO
Romulo Enmark (Sweden)	1995–2001: Chair of Norwegian Research Council on library research (Norway)
Niels W. Lund (Denmark)	1996–1999: Head of Documentation Institute (*Dokvit*) Tromsø (Norway)
Peter Ingwersen (Denmark)	1996–2015: Affiliate Professor at Åbo Akademi University (Finland)
Peter Ingwersen (Denmark)	1999–2002: Guest Professor at Tampere University
Peter Ingwersen (Denmark)	2006–2010: Professor of LIS at Oslo University College in Norway
Peter Ingwersen (Denmark)	2011–2012: Chair of Excellence in LIS at the Department of Library Science and Documentation, Carlos III University Madrid, Spain
Birger Hjørland (Denmark)	1993: doctoral dissertation at the SSLIS in Borås-Gothenburg (Sweden)
Birger Hjørland (Denmark)	2000–2003: Professor at the SSLIS in Borås-Gothenburg (Sweden)

Chapter 5

In Search of an Identity and the Object of a Discipline

In parallel with the official recognition of LIS as an academic field, the 1970–1980s saw the beginning of a quest for a consensual name for the discipline. Different scholars sought to define its subject matter (the ontological question) and the presuppositions and theories on which it founds its methods (the epistemological question). This chapter will recall some of these attempts in a pan-European context with the aim to uncover the network of influences that were at work and the cross-fertilisation of ideas amongst LIS scholars in Europe as they sought to legitimise L(IS). Section 5.1 shows how the evolution in the name of the discipline from "library science", to "documentation" and then to "information science" corresponded to a shift in its object of study : from books to documents and to information.

Sections 5.2 and 5.3 recall previous attempts to find umbrella names for the new discipline which could subsume its different specialties given that the existing terms, Library Science, Documentation and Information Science, were deemed inadequate to encompass all the subfields of LIS. This led us to track how the terms "informatology" and "informatics" were coined, defined and unsuccessfully promoted as umbrella names for the discipline and how they were borrowed by scholars across Europe and further afield, sometimes with similar, sometimes with a differing meanings. Section 5.4 shines the spotlight on how this terminological dilemma has affected the Spanish LIS landscape. Because each suggested name for the discipline brings its own ontological and epistemological presuppositions about the subject of the discipline, section 5.5 looks at how conceptions of LIS have swung periodically from a narrow "physico-processual" conception to a broad "socio-cultural-humanistic" conception, depending on the dominant scientific paradigm at a given time.

5.1. From Library Science to Documentation and to Information Science

The field we now know as LIS has been in the past called "library science" (or library economy and library management), "documentation" and "information science". Each of these names reflected a shift in focus from books to documents and then to information.

The paternity of "library science" [*bibliothekswissenschaft*] is generally attributed to Martin Schrettinger, a German librarian who published a textbook in 1808 in which this term appeared (Buckland, 2017b). By the end of the nineteenth century, bits of library training appeared in vocational schools and later in universities in many western countries. In 1886, a training programme in librarianship was created as a one semester course at the University of Göttingen, Germany. The same year, in the United States, Melvil Dewey, the creator of the Decimal Classification (DC) opened the first "School of Library Economy" at Columbia College which offered a one-year programme and admitted its first students in January of 1887. The choice of the term "library economy" [*bibliotheconomie* in French and *biblioteconomía* in Spanish] instead of "library science" reflected a widely held view that library management was a practical skill and not a science and therefore, one can learn its skills in the same way that one learnt domestic skills or home economics.

The first "library science" programme in the midwestern states was created in the fall of 1893 at the Armour Institute in Chicago (now Illinois Institute of Technology) with two women as sole instructors: Katharine Sharp and May Bennet. To anchor this technical library programme on academic content, Sharp transferred it to the University in Urbana in the fall of 1897 and it became the Illinois State Library School which later evolved into the Graduate School of Library and Information Science (GSLIS) at Champaign-Urbana, one of the oldest and prominent LIS schools in the US[1].

From the beginning of the twentieth century, the term "documentation" began gaining strength as a possible candidate name for the discipline, after all Paul Otlet, the founder of the discipline, had chosen to entitle his celebrated 1934 treatise, "*Traité de documentation. Le livre sur le livre. Théorie et pratique*". The "documentation movement" also crossed the Atlantic and began to permeate the US landscape as the founding of the American Documentation Institute (ADI) in 1937 showed (Williams, 1997, p. 775).

[1]A history of this first program is available on the iSchool of Illinois's website: https://ischool.illinois.edu/our-school/overview/history.

Documentation has since become a versatile and resilient concept of LIS which has survived numerous warnings about its obsolescence and its impending demise in the face of the increasing digitisation of knowledge artefacts.

As information technology emerged in the mid-twentieth century and books ceased to be the sole media for recording knowledge artefacts, the terms "library science" and "documentation" ceded the front stage to "Library and Information Science" (LIS) and later to "Information Science" (IS) in most countries. Buckland (1996, p. 72) described how an intense rivalry emerged between proponents of a "library science" and those of the emerging "information science" in the US during the Cold War era (between 1950 and 1970s). The former was championed by influential social scientists at the Graduate School of Librarianship in Chicago who formed the "Chicago Library school" and vigorously defended a "non-technological, social science oriented paradigm in U.S. library science" (Buckland, 1996, p. 74). After the launching of the Russian satellite Sputnik in 1957, the efficient management of Scientific and Technical Information (STI) became crucial in the US Federal government's efforts to dominate the stars. Several funding programmes were made available for STI projects. A new generation of professors from the sciences embraced information technology as a means of addressing the STI problems that the traditional librarianship was ill-equipped to handle. For many, this was an opportunity to legitimise the field but under the more compelling name of "information science" (IS). The ADI was thus renamed the "American Society for Information Science" (ASIS) in 1968. In 2000, a "T" was added to make it the "American Society for Information Science & Technology" (ASIS&T[2]) in recognition of the importance of technology to the field. Likewise, the bibliographic database *Library Science Abstracts* was renamed the *Library and Information Science Abstracts* in 1969" (Hjørland, 2000, p. 27).

Slowly, library schools in the US became Schools of "Library and Information Science" (SLIS). The change to LIS also marked a return to the historic European documentation roots laid by Paul Otlet, Henri La Fontaine and other European documentalists whose works had been ignored until then in the US not only because of the devastation caused by Second World War but also because they wrote in French and in German (Buckland, 1996, p. 74). Although Paul Otlet and Suzanne Briet used the term documentation and not information science, they are considered by many as the veritable precursors of the field of information science. For

[2]A short history of the ASIST can be found here: https://www.asist.org/about/history/. Accessed on 13 July 2017.

Hjørland (2000, p. 27) Otlet's *Traité de documentation* is one of the "first information science textbooks".

The choice between "library", "information" or "document" as the central concept and object of the discipline is not simply a matter of following trends. As Hjørland (2000, p. 30) rightly pointed out, "The meaning of a scientific concept is always determined by theoretical assumptions. Concepts have no meaning in themselves apart from theories or theoretical assumptions."

However, knowing exactly what theoretical assumptions underpin the use of a concept like "information" has been extremely difficult because of the versatility of the term. Hjørland (2000) discussed at some length the problematic assumptions behind naming the field "information science", i.e., the assumption that the field is dealing with information and not with documents which entails a further assumption that a content (facts, ideas) can be evaluated without taking into account its medium, hence the document from whence it came and without situating it in space and time. These assumptions are closely associated with information retrieval paradigm which dominated the field from the 1960s. Such a physical view of information also echoes Otlet's positivist assumptions that facts can be extracted from books like peas from pods and can standalone in bibliographic systems without their context or source. Otlet accused books of being "messy" and of mixing up facts and non-facts. Hjørland argued that "documentation science", rather than "information science" was a more theoretically-sound name for what the field actually does:

> "The problem that I have raised in this article is whether we should prefer the term 'documentation science' (as recently introduced in Tromsø) or 'information science' (as recently introduced in Copenhagen). I have tried to argue that the conceptions of information, information retrieval and information science are seriously flawed, and that the problems in IS are not just terminological but rooted in problematic theoretical assumptions. This blocking in our field can to a large degree be avoided by changing the object of study from mental phenomena of ideas, facts and opinions, to social phenomena of communication, documents and memory institutions. This is a strong argument for choosing the expression 'documentation science". (Hjørland, 2000, pp. 38–39)

In retrospect, the choice of "document" as the central concept and of "documentation" as name of the department created by Niels W. Lund at Tromsø University in the mid-1990s showed remarkable foresight.

Despite many calls by LIS scholars to replace the word "information" with something else, this has not come to pass. Information is indeed very fashionable and a victim of its own success; thus, it is likely to remain in the name of the field.

Rather than choosing a specific object (document, information, library), it will be more fruitful for the field to focus on informative objects (which can be anything), and on the social contexts in which such objects are produced and can be interpreted, thus adopting a socio-cognitive approach for LIS. Taken separately, each of the above three names "library science", "documentation" and "information science" was perceived to be too restrictive for all the phenomena studied in the field. Hence, efforts were made to find umbrella concepts to subsume the different objects and ideas entailed by these three terms.

5.2. Informatology: Between Information Processing and a Metascience of Information Science

Informatology is one of such umbrella terms suggested in the 1960–1970s as a candidate name for the discipline. Establishing the paternity of this term and its exact meaning has proved elusive. Sections 5.2.1 and 5.2.2 tried to retrace how this term was taken up by different LIS pioneers in different countries, sometimes with the same meanings, at other times with different connotations. At least two major strands of meaning were identified: (5.2.1) informatology as information processing which can be linked to the Swedish LIS scholar Björn Tell; (5.2.2) a second meaning of "informatology" as the metascience of Information Science (IS) defended by Otten and Debons (1970), two IS scholars in the United States. Section 5.2.3 recounts efforts by Jean Meyriat to legitimise information-documentation as a scientific field in France by promoting the two concepts of "documentology" and "informatology".

5.2.1. Informatology as Information Processing: The Swedish Trail

Our first encounter with "informatology" was in the writings of the Yugoslavian pioneer of LIS, Božo Težak who chose this term as the name of the new discipline he was establishing in Yugoslavia and of the journal *Informatologia Yugoslavica* he founded in 1969 as a communication platform for the nascent discipline:

> The very name of INFORMATOLOGY is meant to designate theory and practice of emission, transmission, accumulation, selection and

> absorption of information (the so-called e-t-ac-s-a complex[3]) and is wider and more exact than terms such as informatics, information sciences and services. (Težak, 1969a)

Težak (1969a, II) further specified that "INFORMATOLOGY first appeared in 1962 (1)". His reference section (1) contained a rather vague citation to "P. Atherton, *Amer. Docum.* 16 (1965) 126".[4] Further search revealed that this reference referred to a one page "Letter to the Editor" by Pauline Atherton, published in the April issue of *American Documentation* informing the readers about a new 100-hour course named "informatology" which had been created in the autumn at the Royal Institute of Technology (KTH), Stockholm in Sweden. According to Atherton[5]:

> The title of this 100-hour course has its origins in Sweden. As far as Björn Tell, librarian at the KTH, knows, informatology was first mentioned in a lecture given on March 8 1962. The intention was to avoid the term documentation which connotes attention to documents. Now that the emphasis in our field is on information, it was considered appropriate to coin a term with that connotation.

Atherton further specified that the new course was "arranged by the Swedish National Committee on Documentation under the general direction of Dr Tell".

The wording of Atherton's *Letter* was rather vague. It did not explicitly state that Björn Tell coined the term in 1962 nor that he was the author of the lecture. Furthermore, the definition of "informatology" given in her *Letter* was confusing and appeared to be a poor attempt at paraphrasing the original definition. Hence, further search was needed to find the publication where informatology would have been coined and defined for the first time.

Our search led us to a 2003 interview of Börje Langefors in which he evoked a meeting with Kjell Samuelsson[6] and Björn Tell[7] around 1965 when Langefors was trying to create a research group in information processing. The three men tried to find an "interesting name" for computer-aided

[3]This is a central concept in Težak's understanding of information and communication processes to which we will return later.

[4]*American Documentation* is the ancestor of *Journal of the American Society for Information Science* (JASIS) and of the current *Journal of the American Society for Information Science & Technology* (JASIST).

[5]More information about Pauline Atherton can be found on ASIST's page of information science pioneers: https://www.asist.org/pioneers/pauline-atherton-cochrane/.

[6]Samuelsson was a professor of Informatics and Systems Science at Stockholm University.

[7]Björn Tell was chief librarian at Royal Institute of Technology (KTH), Stockholm.

problem solving which was of particular interest to Langefors. They decided on "information research" or "informatology" later abbreviated to "infology" as proposed by So Sundgren[8]".

Once again, the wording of this interview report remained vague and did not lift the ambiguity surrounding the origin of the term "informatology" beyond all reasonable doubt. Moreover, the date of the meeting remained imprecise.

Further search uncovered a book entitled *ICT for the people. 40 years of academic development in Stockholm* published in 2006 which was a compendium of accomplishments in the field of information technology in Sweden. In this book was an article by Kjell Samuelson (2006, p. 196) in which he wrote about a new postgraduate course designed by a transdisciplinary group on information retrieval, systems and classification research formed in the early 1960s, of which he was a member as well as a certain B.V. Tell:

> our study group in 1963-64 produced the outline of a postgraduate course on the fundamentals of "Scientific Information Transfer and Retrieval". It was running over an 8-month period with 20 credit units. The course was given at the Royal Institute of Technology, with financial support by the Council for Applied Research (Tell, 1965). Course modules treated topics such as information transmission and retrieval (IR), systems analysis, cybernetic systems, neuronets, automata, selforganization, classification & clustering, coding & redundance. A new speculative module was coined "informatology" to denote a research field combining intuitive, heuristic procedures with augmentation of human intellect and reasoning by computers and other media. (Engelbart, 1963)[9]

Unsurprisingly, Samuelson's conception of informatology bears the mark of his cybernetics and systems theory background and of his preoccupation with computer analysis of data and artificial intelligence. More importantly for our quest, Samuelson cited a 1965 article by Björn Tell published in the proceedings of a Congress held jointly the same year by the FID and the American Documentation Institute (ADI) in Washington DC.

---

[8]Börje Langefors is an Emeritus Professor at the Department of Computer and Systems Science, University of Stockholm and Royal Institute of Technology (KTH), Stockholm. His interview is published in Bubenko J., Impagliazzo J. & A. Solvberg (Eds.), the Proceedings of the First Workshop on History of Nordic Computing, June 16–18 2003, Trondheim, Norway, p. 14. Available online at: https://books.google.hr/books?id=HGTgBwAAQBAJ&pg=PA14&lpg=PA14&dq=bjorn+tell+swedish+librarian+informatology&source=bl&ots=8-. We are indebted to Tatjana Aparac-Jelusic for sending us this link.

[9]The references Samuelson gave are respectively Tell (1965) and Engelbart (1963).

When we finally got hold of this 1965 article, Björn Tell argued in it about the need to support advances in science and technology with an efficient training programme in documentation. He explicitly used the terms "documentation" for the field and "documentalists" as an umbrella term for "special librarians, information specialists and technical literature analysts, etc.". He then described an eight-month long new experimental course delivered at the Royal Institute of Technology in Stockholm in the Fall of 1964 in order to train documentalists for the efficient management of scientific and technical information in science and the industry.

This new course, he further specified, was being crafted as a result of a 1962 report of a committee appointed by the Royal Academy of Engineering Sciences which concluded that "there was a need for about 30 qualified documentalists annually up to the end of 1970". This committee proposed a "comprehensive training programme" for documentalists that covered all university levels, from graduate to doctorate (Tell, 1965, p. 41). The course content was aimed at not only acquiring traditional cataloguing skills but also computing skills to enable trainees to efficiently manage scientific and technical information for various organisations" using the most up-to-date technology. Because emphasis was laid on "information" and "the analytical aspects in the use of computers for information handling", it was decided to avoid the term documentation which "connotes attention to documents", hence the choice of the word "informatology" (Tell, 1965, p. 43). At the time, digital documents had not yet become commonplace and "document" and "documentation" referred only to the print medium. Visiting professors were invited to give lectures in this course one of whom was the French pioneer of automatic indexing, Jean Claude Gardin.

From the description of this course, it is the same as the one referred to in Samuelson (2006) although Tell slightly altered the title to "Transmission of Scientific Information" and wrote that it "was intended as an introduction to "informatology" (Tell, 1965, p. 44). Hence, one can deduce that Björn Tell ran this course with the transdisciplinary research group mentioned in Samuelson (2006). Also, from the presentation of the course, "informatological aspects" is but one aspect of the entire course on Scientific Information Transmission.

As to what the term *informatology* meant, Tell wrote that:

> The word "informatology[10]" was coined to denote the research field within information processing which combines intuitive and algorithmic procedures. Computers are used merely as a supplement to the human intellect. Informatology will handle both those subsets of the

[10]In italics for emphasis in the original text.

> information process which are not dependent on intuition or knowledge and those in which it is a prerequisite that intuitive procedures are included. The purpose is to arrange the procedures in such a way that the intuitive subsets are distinguished from those which might be formalized, and that the two kinds of subsets are regarded as well-defined components in a total system.

The wording of this paragraph still remained frustratingly elusive. Who coined informatology: Björn Tell alone? Björn Tell and his fellow course designers? Was it "coined" during the 1962 lecture? If so by whom? Or was it in the 1962 Committee Report ordered by the Royal Academy of Engineering Sciences that recommended that the new course be "considered" by the Swedish National Committee for Documentation before being tried at the Royal Institute of Technology in the Fall of 1964?

Concerning what informatology meant, Tell's paragraph offered more clarity. Steeped in computer science and artificial intelligence, informatology was perceived as the art of deciphering instances when human intellect and input is needed to perform a given information processing task and instances where it is not required and the tasks can be wholly automated. Informatology does not deal with either one of these cases but with both.

Logically, Tell envisioned that people endowed with such expertise will be called "informatologists". Their task will be to formalise intellectual activities such that computers can be used to simulate them:

> The treatment of subsets of intellectual activities which do not depend on intuition is in the domain of the "informatologist". He studies an activity like classification, which generally do not presuppose the use of computers or formal logical methods, in order to analyse its intuitive character and ascertain whether this activity (a series of intuitive sets) can be formalised and defined as subsets of the larger system, which would then employ computers. (Tell, 1965, pp. 43–44)

The dominance of artificial intelligence in the conception of informatology is further revealed in the subjects comprising the "Informatological aspects" of the course:

> codes and error correction etc,
> Neural network,
> Self organization and automata
> Uncertainty theories, "redundant information", etc;
> Learning and problem solving, etc.
> Theory of automata,
> Linguistic problems in information retrieval. (Tell, 1965, p. 44)

The tasks Tell assigned to the informatologist are similar to those of the modern day programmer and computer scientist. Tell's informatologist is therefore a translator of human needs into formal machine language:

> The hope is that identifying problems of this kind will have a potential value for the informatologist when he proceeds to research. He will also be better prepared when new systems and machines are introduced. Even if the informatologist seems to be biased towards information processing and development of know-how of computers, he should also know when not to use a computer. He should be equipped by his training to make decisions about new methods and tools to be used for information storage and retrieval in the same way as a research worker in a firm will influence the decisions on new research project. (Tell, 1965, pp. 43–44)

At the time of publication of his article (Fall of 1965), permission to continue the course and integrate it into Stockholm University was not granted by the Swedish Ministry of Education.

5.2.2. *Informatology as a Metascience of Information: The American Trail*

The term "informatology" resurfaced again in the title of an article by Klaus Otten and Anthony Debons in 1970 but with an entirely different meaning. These authors argued that information is a fundamental and universal phenomenon similar to matter and energy[11] and that the operations performed on information require fundamental phenomena. They further argued that "the body of knowledge describing these phenomena and relations will evolve as the subject of a new science which will unify existing concepts while "deriving formalistic descriptions" (Otten & Debons, 1970, p. 89). Their conception of IS is thus that of a natural science which needed unifying theory and formal laws. According to them, the need for a metascience arises when existing bodies of knowledge become hyperspecialised and their focus narrower and narrower.

They then proposed "informatology" as the name of a metascience of information whose role it is to unify the theories and concepts of the contributing sciences using a common language that subsumes them:

> As Metamathematics evolved in response to the divergence and growths of specialized mathematical disciplines, similarly, we

[11]This is the contrary of Wiener's assertion that "Information is information, not matter or energy", Cybernetics, 2nd edition, p. 132.

> anticipate the evolution of a metascience of information in response to the need for a critical reevaluation of the foundation upon which many information disciplines and technologies are based. This anticipated metascience can be viewed as the science of information (or informatology). Informatology can be defined as the study of the fundamental principles underlying the structure and use of information. [...] Thus, the metascience of information has two focal points: the phenomena of information and man's relation to the phenomena. (Otten & Debons, 1970, p. 92)

The authors further specified that their conception of a metascience of information is not meant to be an all-inclusive science as certain definitions have tried to be by saying that IS "deals with all aspects of information". For them, "the metascience of information as a very specific science, concerned only with the foundations of information-related sciences and technologies and not concerned with the content of these specialized disciplines. Any claim for all-inclusiveness would lead to superficiality and therefore would not serve any purpose" (Otten & Debons, 1970, p. 94).

5.2.3. Documentology and Informatology: The French Trail

In the years following the official recognition of Information and Communication Sciences (ICS) in France, Jean Meyriat (1981a) attempted to define the object of the IS branch by suggesting that "documentology" be used in *lieu* of documentation and "informatology" in *lieu* of "information science".

- *Documentology as the science of documents*
  Meyriat first recalled that document is the medium on which information is fixed and that one cannot define the former without being interested in its content which is really the object of every documentary process. He then suggested that the field of documentation should be renamed Documentology [*documentologie*].

  Meyriat (1981a, p. 59) stated that he borrowed the term *documentologie* from Robert Escarpit, another founding father of ICS who had used it in his 1976 book "General theory of information and communication" [*Théorie générale de l'information et de la communication*], considered as one of the earliest attempts to articulate a general unifying theory for the nascent interdiscipline. Escarpit also apparently employed *documentologie* in an earlier book entitled *L'écrit et la Communication* published in 1973 (Presses Universitaires de France, 56) in the sense of "the science of documentary techniques" [*science des techniques documentaires*]. For Meyriat, "documentology" was the general science of all the specific fields devoted to scientific discourse on different types of documents such as "bibliology, iconology, filmology, discology, etc" [*bibliologie, iconologie, filmologie, discologie*]. Meyriat noted that the term was equally in use in Spain but gave no further details about who may have used it. We found that Sanchez Belda had suggested the term "*documentologia*" in 1962 as a possible name for the discipline in Spain (see section 5.4 infra). Meyriat further observed that documentology was not included in the Terminology of Documentation published in five languages in 1976 by the UNESCO nor in the French national standard NF Z40-001 on "Vocabulary of Information and Documentation" [*Vocabulaire de l'information et de la documentation*] which gives the official nomenclature of the field.

- *Informatology as the science of information processing*
  Having defined "documentology" as the "general science that studies the different types of documents", Meyriat naturally turned his attention to the content of documents, i.e., information. Meyriat suggested that "*informatologie*" (informatology) be used to name the science of information. Hence, if the focus of documentology was on the medium (document), that of informatology was on the content (information):

  > informatology seeks to give an account of all systems of transfer of useful information, whether primary or secondary. It studies also information systems that do not have recourse to documentary medium. In particular, it focuses its attention on the processing of information itself, regardless of its medium. (Meyriat, 1981b, p. 18)

Meyriat's informatology is a less "machine-reliant" version of information handling than Tell's and Težak's. While his informatology has as focus information processing and transfer, the computer programming dimension is not as prominent. Meyriat proposed to name 'informatologist' [*informatologue*], a person who makes a scientific discourse on information and "informatist" [*informatiste*], the person that implements the results of this discipline and practises the transfer of information. Hence Meyriat's "informatologist" does not quite equate to Björn Tell's where an informatologist was synonymous with a computer programmer-analyst. The two were using the same terms in near synonymous but not quite the same sense. Tell's informatology and informatologist is more computer and AI oriented than Meyriat's.

Also, observe that Meyriat's documentology has a much broader scope than his use of informatology. Whereas his "documentology" encompassed all types of documents, his "informatology" was only concerned with the information useful for teaching, research and in the industry. By association, the field of IS was to be only concerned with scientific and technical information (Meyriat, 1981a, p. 62). Meyriat considered "information science" too restrictive a term, a view echoed later in Hjørland (2000, pp. 38–39) who also preferred documentation to information science as a better name for the discipline (see section 5.1 supra). Meyriat (1981a) considered "informatology" and "documentology" to be the two branches of a field which he suggested should be called *Science de l'information et de ses supports* (the science of information and of its media).

Meyriat's *oeuvre* of terminological innovation was not successful in France although it received some resonance in the North African LIS community (Tunisia, Algeria, Morocco) where the term "*informatiste*" is often used to refer to information professionals.

Also, Meyriat and Escarpit were not the first in the francophone zone to promote the use of these terms ending in –logy (*–logie*) to name the different subfields of LIS. The first chapter of Paul Otlet's celebrated 1934 *Traité de documentation* was entitled "*La bibliologie ou documentologie*". While Otlet treated both terms as quasi synonyms, he used "*bibliologie*" in the text of the chapter. Pagès" 1948 paper which foreshadowed Briet's famed "antelope-as-document" example was entitled "*Transformations documentaires et milieu culturel (Essai de documentologie)*". It appears that the choice of *documentologie* at the time was logical since "*bibliologie*" (bibliology) was also promoted (unsuccessfully), as the name of the field devoted to the study of the printed book ("material bibliography"). It followed then that the study of documents (including books), ought to be called "documentology". This attempt also proved to be equally unsuccessful.

Although Meyriat (1981) did not cite the prior use of informatology by other European predecessors, one cannot rule out the possibility that he was aware of them. In our interview with Neva Tudor-Silovic who was Božo Težak's assistant, she stated that she had met Jean Meyriat and that their last meeting took place in 1984 at a council meeting of the FID. Meyriat seemed to be well integrated into the network of European pioneers of his time. He lived in Spain during the Second World War and had interacted with some of the Spanish IS scholars.

Interestingly, the bid to designate the discipline with the *–logy* suffixes met with success in Poland. After several decades of oscillating between different names for the discipline amongst which "librarianship, documentation, information science, informatology, informology, informatorics, informatics, infoscientistics, bibliotronics" (Cisek, 2002, p. 45; Dembowska, 1965, p. 25), information science gained prominence at the turn of the 20th and 21st centuries as the preferred name for higher education programmes. However, in 2011, a reversal of fortune occurred. By ministerial decree, "bibliology and informatology" was chosen as the new name for the discipline in Poland. The rationale behind the choice of not one, but two terms with the *–logy* suffix was to convey the idea of an interdisciplinary field, concerned not only with the scientific study of the printed book but also encompassing all the subfields one usually finds under information science programmes such as information behaviour, reading literacy, uses of ICT, book history, tegumentology[12], information policy and ethics, etc[13].

5.2.4. Summary of the Informatology Track

To sum up our findings, it appears that informatology has been used in two very different and somewhat opposing senses. First, a narrow sense which is the one adopted by Björn Tell and his colleagues at the Royal Institute of Technology in Sweden, taken up by Pauline Atherton (1965), Božo Težak (1969a) and Jean Meyriat (1981a,1981b), *albeit* with a lighter information processing component. Secondly, a broader sense by Otten and Debons (1970) in the US.

In its first narrow sense, informatology is concerned with man-machine symbiosis and draws on artificial intelligence to determine what aspects of a knowledge rich task can be formalised and programmed.

The *informatologist* is therefore a specialist in data processing methods and techniques. This is also the sense in which Samuelson and Langefors refer to *informatology* or *infology*. This narrow and processual conception of informatology takes its origins from the computer paradigm. It would be synonymous with a narrow and restrictive sense of informatics today.

In its second broader sense as used by Otten and Debons (1970), informatology refers to the study of the theories and foundations of information science and thus to a metascience of IS. The focus is on the quest for unifying theories and principles underlying the study of information. However,

[12]Having to do with study of the history of book binding. The Institute of Science Information and Book Studies of Nicolaus Copernicus University in Toruń organised the First National Bookbinding Studies Conference "Polish Bookbinding Studies Today" on 26th–27th of June 2014. See https://wydawnictwo.umk.pl/pl/products/3562/tegumentologia-polska-dzisiaj-polish-bookbinding-studies-today.

[13]We are grateful to Zbigniew Gruszka for providing us with a short summary of the emergence of higher education programs in LIS in Poland and on the terminological oscillations surrounding the name the discipline. We were unable to investigate further the sources he cited in his summary which were all in Polish. Hence, a more detailed study of the Polish LIS history is needed.

proposals to make a field a metascience of another field are doomed to fail because as Tercar (2000) observed, disciplines are in interaction with one another and are in a horizontal relation whereas a metascience suggests a vertical or hierarchical relation.

5.3. Informatics and the Information Processing Paradigm

Around the same time that informatology was being promoted as an umbrella term for IS, informatics was also being promoted for the same purpose. When searching for a name for his new journal and for the emerging discipline he was creating, the Yugoslavian pioneer Božo Težak considered three choices: informatology, information sciences and informatics. Concerning the possibility of choosing informatics, he wrote:

> The term INFORMATICS was almost simultaneously or somewhat later defined as "a new scientific discipline dealing with structure and properties (but not with concrete contents) of scientific activities, their theory, history, methodology and organization." (Težak, 1969a, II)[14]

Two years later, Težak (1971) added that "informatics is the common term for the field of information sciences and services while informatology should comprise the whole complex of *emission, transmission, accumulation, selection and absorption of information*".

Thus, for Težak, informatics is synonymous with the entire field of IS whereas its original authors, as we will show hereafter, gave it a more restricted definition. Note also that Težak's broad conception of informatics is more in line with Otten and Debon's broad conception of informatology. Concerning the paternity of informatics, Sercar (2010, p. 282) brought more details to Težak's vague account:

> The term informatics was first used by Ph. Dreyfus in 1962, and the same year B. Tell introduced the term "informatology" in Sweden. The term "informatology" was also supported by P. Atherton (1965). Starting with 1965, A.I. Mihaijlov, A.I. Cerni and R.S Giljarevski spoke in favour of the term "informatics".

[14]Težak's referencing style is quite cryptic. The (2) juxtaposed beside this definition led to a footnote listing the following authors: "Ph Dreyfus, Gestion (1962) 24o; J.G. Dorfman, NTI 12 (1966) 46; A.I. Mihaijlov, A.I. Cernyj, R.S. Giljarevskij, NTI 12 (1966) 35; FID New Bull. 17 (1967) 7o; cit.: Informatika 1 (1967) sep." Težak's (1969, p. III). Težak did not make precise which of them was the author of the definition he cited. It appears to be a compilation of all the authors who had used "informatics" as a possible substitute for information science.

This passage pointed to two trails: the French and the Russian.

5.3.1. *Informatics as Information Processing or Computer Science: The French Trail*

An Internet search showed that the "Ph Dreyfus" referred to in the Težak's quotation was most likely the French computer science pioneer Philippe Dreyfus who was Director of the National Computing Center of the computer company Bull in the 1950s.[15] Dreyfus was credited with being the first to translate the term *informatik*, coined apparently by the German engineer Karl Steinbuch in 1957, into its French version in 1962. We were however unable to locate the first German version in any publication and to muddy the waters, Samuelsson (2006, p. 197) attributed the first appearance of the term to the French *informatique* and not to the German *informatik*:

> The name informatics is adapted from the french "informatique" and frequently used in translations to European languages. (Informatik, informatikka)

If indeed the French "*informatique*" is the first appearance of "informatics", it has a broader sense since "*informatique*" refers to computer science whereas the English "informatics" was given a more restricted meaning in Samuelson (2006, p. 197):

> Informatics is Information Science and Technology (IT) for multidirectional information flows through and between living systems: individuals, groups and organizations, foremost international knowledge transfer and retrieval, in behavioral harmony with matter-energy flows.

It is probable that in those early years of the emergence of computer science, its frontiers were still unclear and it was not as broad a field as it is today. Dreyfus's first use of *informatique* in 1962 may therefore have been in a more restricted sense of "computer handling of data" which would have been closer to its English variant informatics at the same time. This is only a conjecture at this stage for which we lack written evidence.

---

[15]We are indebted to Thomas Dousa, librarian at the Chicago University Library for helping us unearth P. Atherton's letter and resolving the mystery around her second name Pauline Cochrane (née Atherton) and also for ferreting out the likelihood of "Ph Dreyfus" being "Philippe Dreyfus". A short wikipedia page is dedicated to Philippe Dreyfus at https://fr.wikipedia.org/wiki/Philippe_Dreyfus. Also, mention is made of his use of the term "informatique" in the "*The book of Informatics*" published in 2006 by John Gammott, Valerie Hobbs and Diarmuid Pigott, p. 2.

More clarity was brought to our quest for the origins of informatics from unexpected quarters. In his 1992 doctoral thesis written in Catalan, Ernest Abadal (1992, p. 96) had also tracked the origins of the different names suggested for the discipline worldwide. His account corroborated Samuelson's but offered a more complete picture of the origins and uses of informatics. Abadal confirmed that Ph. Dreyfus coined the French version *informatique* as a contraction of *information* + *automatique* in 1962 at a Meeting of the "French Association for Computing and Information Processing" [*Association Française de Calcul et du Traitement de l'Information* (AFCALTI)] to name the new domain related to computers. Dreyfus was also credited with giving the linguistic versions of *informatique* in five other languages. This reinforces our conjecture that Dreyfus's use of the term in 1962 may have been in a more restricted sense than the sense of "*informatique*" today.

5.3.2. Informatics as the Science of Scientific Information: The Russian Trail

According to Abadal (1992, p. 96), the origins of the Russian[16] version of the term *informatika* can be retraced to a 1965 handbook entitled *Fundamentals of Scientific Information*[17] written by Mikhailov, Txernii and Giljarevski. This was three years after Dreyfus's use of the French version. Mikhailov had studied mechanical engineering and worked in the aeronautical industry before being interested in Scientific and Technical Information (STI). He went on to take part in the creation of the Russian agency VINITI [*Vserossijsky Institut Nauchnoi I Tekhnicheskoy Informatsii]* in charge of STI and became its director in 1956, a position he held until his death[18]. Mikhailov and his co-authors were looking for a less controversial term than information with which to name the new discipline. They considered and rejected the term "scientific information" used in the title of their handbook because it referred more to an object of study than to an area of knowledge. They equally rejected the term "documentation" because it appeared to restrict scientific information activities to documents only.

[16]At the time, Russia was the center and the seat of power of the Soviet Union (also known as Union of Soviet Socialist Republics (USSR)) which existed between 1922 and 1991 and comprised 15 communist Republics. More information can be found on its Wikipedia page: https://en.wikipedia.org/wiki/Soviet_Union.

[17]Original title in Croatian was "*Osnovi Nauknoi informatsii*". (Moscou, Nauka, 1965, cited in Abadal, 1992, p. 96).

[18]According to his brief biography of Mikhailov, available here https://en.wikipedia.org/wiki/Alexander_Ivanovich_Mikhailov.

Therefore, a brand new term had to be found. At this point, one of the reviewers of their handbook, a certain J.G. Dorfman suggested the term "informatics" [*informacija*[19]]:

> It would be more reasonable to use *informatics* as the name of a science which deals with the basic ideas, methods and means of collecting, processing, storing, retrieving and disseminating any one type of information. (quoted by Mikhailov, 1967, p. 72 de Dorfman, J.G. "Review of the book …", *Naucno-Teknicekaja informacija*[20], 1966, 7, 46 cited in Abadal (1992, p. 96)

In 1966, Mikhailov and his colleagues published an article entitled "*Informatika: new name to the Scientific Information Theory*" in the journal *Naucno-Tekniceskaja Informacija*. This article was republished in English in 1967 in the FID journal.

At the 33rd general conference of the FID which took place in Tokyo in September of 1967, Mikhailov defended the use of the neologism in his paper. In 1968, the second edition of his manual was published under the title "*Fundamentals of Informatics*" [*Osnovi Informatiki*] which became the first treaty bearing this term.

In 1971, the UNESCO and the FID asked Mikhailov and Giljarevski to publish a guide entitled "*An introductory course on Informatics/ Documentation*[21]". The explanatory note on the cover page however read, "*This draft text for an introductory text in documentation science is reproduced as a working document in the English language submitted by the FID*". The hesitation between "informatics" and "documentation" in the title indicated that the authors were unsure about the acceptance of the former and considered it more prudent to accompany it with the more widely used and better understood "documentation". As to what informatics meant, the authors wrote:

> Informatics is a scientific discipline which investigates into the structure and properties (and not the specific contents) of scientific information as well as the regularity of scientific information work, its theory, history, methodology and organization.
>
> The purpose of informatics consists in developing optimal methods and means of presentation (recording), collection, analytical-synthetic

[19]This is the Lithuanian version. Lithuania was one of the Republics of the Soviet Union.

[20]"*Naucno-Teknicekaja informacija*" is the latin (Bosnian) transliteration of the cyrillic Russian alphabet of *Научно-ТехническаяИнформация* which means "Scientific and technical information".

[21]This document is available online at https://files.eric.ed.gov/fulltext/ED060875.pdf. Accessed 11/12/2018.

> processing, storage, retrieval and dissemination of scientific information. Informatics deals with logical (semantic) information, but is not involved in qualitative estimation of this information. Such an estimation can be carried out only by specialists in the particular fields of scientific or practical activity. (Mikhailov & Giljarevski, 1971, p. 14)

The first part of the definition was the one reproduced in Težak (1969a). The emphasis on the structure, properties and regularity suggest that the focus is on identifying the laws that govern scientific information as a system. In this sense, informatics encompasses what has become the metrics fields (biblio/sciento/infor-metrics).

The second part of the definition echoes Harold Borko's 1968 definition for IS and Težak's conception of informatology by its processual dimension. However, Mikhailov and Giljarevski's informatics appears to be restricted to scientific information only and does not appear to be concerned with the other types of information.

The fact that Mikhailov was able to leverage the communication channels of the FID and UNESCO in order to promote his new concept is explained by his role within these organisations. He was at the time coordinator of FID's committee on "Research on Theoretical Basis of Information" (FID/RI) and Vice Director of this organisation between 1969 and 1976 and again between 1981 and 1988. The Russian agency VINITI of which he was a founding member oversaw the co-edition of numerous FID bibliographies (Abadal, 1992, p. 99).

5.3.3. Summary of Informatics Trails

The promotion of informatics as a possible umbrella name for IS found some resonance in the name of several journals that appeared in the 1970s. In 1969, "*Informatik: Information und Documentation in Wirtschaft, Wissenschaft und Technik*[22] appeared in the German Democratic Republic. In 1967, *Informatika: Casopis za teoriju i praksu naucnih, technickih i ekonomiskih informacija i dokumentacije*[23] appeared in Yugoslavia. The same year (1967), *Informatika* appeared in Hungary and in 1970, the journal *Naucno-Teknicskaja Informacija* was renamed *Informatika* in the Soviet Union (USSR).

[22]German for "Informatics: Information and Documentation for Economy, Science and Technology".

[23]Croation for "Informatics: Casopis for the theory and practice of scientific, technical and economic information and documentation".

Despite all these efforts to install *informatics*, the international LIS community remained unconvinced. A plausible explanation may be the ambiguities of its different linguistic versions which explains why it was not successful in supplanting information science as the name of the discipline (although the same reason of ambiguity has not stopped "information" from being adopted as the name of the discipline):

- the French *informatique* coined in 1962 by Ph. Dreyfus has come to signify the entire field of computer science;
- the Russian equivalent *informatika* had a much narrower scope and was concerned only with scientific information processing; and
- the English equivalent *informatics* refers to the computer handling of data, hence to data processing. Informatics is that part of information science that draws on computer science and statistics to build information systems for specific application areas (bioinformatics, medical informatics, etc).

5.4. Terminological Dilemma in the Hispanic Zone: Between Francophone and Anglophone Influence

Like many other countries, Hispanic countries have been battling with the terminological dilemma which has dogged the LIS field since its emergence. Abadal (1992, p. 260) carried out an inventory of the different names suggested by LIS pioneers in Spain to name the discipline and the contexts in which they were proposed of which we summarise the salient points. One of the pioneers, Lasso de la Vega proposed naming the discipline "*Documentation*" [*Documentació*] in 1954. Garcia Morales suggested "documentalisme" in 1962 while Sanchez Belda proposed "Documentology" [*Documentologia*] the same year. By the end of the 1960s, a change in terminology became perceptible: in 1968, Penna and Sanchez Belda suggested calling the discipline "Scientific and Technical Information" [*Informacio Científica i Tècnica*]. This was the same year that Harold Borko's definition of information science was published in North America. Coincidence? In 1969, Roger Riviere suggested "Scientific Documentation" [*Documentació científica*] while "Informatics" [*Informàtica*] made its appearance on the Spanish scene in 1969, suggested by Mut Calafell.

In 1971, Vicentini Abner Lelis Correia published "From library management to informatics. Evolution of the information concept" [*De la biblioteconomía a la Informática. Evolución del concepto de Información*] in which he offered a systematic review of the definitions and theories of documentation. A year before (1970), Pérez Álvarez-Ossorio suggested that the field be named "Information and Documentation" and also that documentation be integrated into the tertiary sector as a branch and service of scientific research.

The end of the 1970s witnessed a return to the documentation root as shown in "Science of documentation" [*Ciència de la documentació*] by Lopez Yepes in 1978 and "Documentation sciences" [*Ciències de la documentació*] (in its the plural form) by Emilia Currás in 1982.

Their conception of "informative documentation" appears to refer to the documentary processes applied to different areas of activities, such as scientific communication, journalism, audiovisual communication, advertising and public relations.

In his 1995 book "Documentation as discipline" [*La documentación como disciplina*], Lopez Yepes identified 23 of the most used terms to name the discipline and provided a chronological chart of the denominations between 1947 and 1994. In a later publication, he identified several universes of meaning of the term documentation[24]:

1. *Documentation* can designate *a set of fields* (Documentation Sciences) whose object is the study of the document as a source of information to obtain new information or to make decisions.
2. Documentation can be understood as a *documentary discipline* or "General Documentation" that studies the context (systems, networks and information centres) where documental operations take place.
3. Documentation becomes an *instrumental discipline* when it deals with the document as a potential and effective source to obtain new information that serves the growth and transmission of knowledge.
4. Documentation can be *normative* because it affects the way ideas are transmitted.
5. Documentation is *general* or *autonomous* in its service to knowledge which places it within the field of science of sciences.
6. Documentation is equally *specialised* because it becomes operational in the context of service to a particular branch of knowledge or social activity.
7. Finally, it is *scientific* when it provides support to the process of obtaining new scientific knowledge.
8. Documentation takes as the purpose of study *an informative process* of a special nature called "informative and documentary process". This process is based on the preservation and retention of previous informative messages that are retrieved and subsequently transformed into documentary messages. This process has a teleological component: the receiver uses these messages as source to create a new informative message which

[24]This account is based on a translation of the original article in Spanish.

will enter again into this continuous and cyclic process. (Lopez Yepes, 2001, p. 263)

When stripped to their bare bones, Yepes's different meanings of documentation appear to boil down to their pragmatic and utilitarian essence. He talks of an instrumental discipline and of a normative process. Documentation is seen by him as a process by which knowledge in a field is "utilised" in order to produce new knowledge and so on.

Lopez Yepes and his colleagues at the Faculty of Information Science (in reality the Faculty of Journalism and media studied) at Complutense University in Madrid tried to promote the concept of "informative documentation" in the late 1970s which they defined as "a two-headed discipline, aimed at studying the process through which new scientific knowledge is obtained in the Social Communication Research (Documentation of the Information Science) and the work inside the informative and communicative companies (Documentation of the informative work)". The author observed that their definition is tautological "since all kind of documentation is considered as information, although the opposite is not true". (Yepes, 2001, p. 269).

From the 1990s, "Documentation" [*Documentación*] and Library management [*Biblioteconomía*] appear to have gained acceptance in Spain as the umbrella term for all the subfields of LIS (information science, librarianship, archival studies and museology) for reasons we have already explained in chapter 4: on the one hand, the term "information science" had already been appropriated by media studies and journalism and therefore was no longer available; on the other hand, Spanish LIS scholars were greatly influenced by the francophone documentation tradition (Abadal, 1992, p. 260). However, Yepes (2001) observed that other terms like "information management" and "knowledge organization" are now being used to designate other aspects of LIS not captured by documentation. Lately, the terms "documentation" and "librarianship" appear to be falling into disfavour due to lack of public recognition both from the government and the job market. This has led the LIS unit in Carlos Madrid III university to rename its bachelor's programme in 2017 as "BA in Management of Information and Digital Content[25]".

Lopès Yepes and Sabelli (2013) observed that there is not yet a single consensual name on which everyone agrees for the discipline in the Ibero-American countries (Latin America, Spain and Portugal). Their survey found that the field is designated by not fewer than eighteen different

[25]See https://www.uc3m.es/ss/Satellite/Grado/en/Detalle/Estudio_C/1371224922548/1371212987094/Bachelor_s_Degree_in_Management_of_Information_and_Digital_Contents#presentacion.

names resulting from a combination of its three core concepts "library, documentation and information":

(1) *Bibliotecología y Ciencia de la Información (Library Science and Information Science)*
(2) *Bibliotecología y Ciencias de la Información (Library Science and Information Sciences)*
(3) *Bibliotecología y Documentación (Library Science and documentation)*
(4) *Bibliotecología y Estudios de Información (Library Science and Information Studies)*
(5) *Bibliotecología y Estudios de la Información (Library Science and Studies of Information)*
(6) *Bibliotecología, Documentación e Información (Library Science, documentation and information)*
(7) *Biblioteconomía (Librarianship)*
(8) *Biblioteconomía con habilitación en Gestión de la Información (Librarianship with qualification in information management)*
(9) *Biblioteconomía e Información (Librarianship and information)*
(10) *Biblioteconomía y Documentación (Librarianship and documentation)*
(11) *Ciencia de la Información (Information Science)*
(12) *Ciencia de la Información con habilitación en Biblioteconomía (Information Science with Qualification in Librarianship)*
(13) *Ciencia de la Información y Documentación (Information Sciences and documentation)*
(14) *Ciencias bibliotecarias y de la Información (Library and Information Sciences)*
(15) *Ciencias de la Información (Information Sciences)*
(16) *Ciencias de la Información documental (Documentary Information Sciences)*
(17) *Ciencias y Tecnologías de la Información (Information Sciences and Technologies)*
(18) *Documentación (Documentation)*.

The authors wondered if this lack of consensus around a single name was not synonymous with the absence of agreement on the signification of these concepts and their theoretical foundations.

They noticed a marked preference for "Information Science" in Latin America where it is rarer to encounter "Documentation" as the name of a LIS programme.

They also noted that in the 1980–1990s, Latin and in particular South America displayed a more integrative attitude to LIS. "In contrast, in Europe and the United States, the trend in higher education was for Library Science, Documentation and Information Science to grow apart"

(Octavio Rojas[26] cited in Lopes Yepes and Sabelli, 2013). A possible explanation was that developing countries which had not yet experienced the same level of advances in information technology adopted a more integrative approach to LIS than industrialised countries where the rise of computer science with its promises of automation of information tasks led to proponents of an "Information science" distinguishing it from "Librarianship".

5.5. The Dance between Narrow and Broad Conceptions of Information Science

This last section dwells on attempts to define the object of the discipline and how that shaped the European conceptions of IS. The pendulum has swung from a narrow "physical-processual" conception (section 5.5.1) to a broad "socio-cultural and humanistic" conception (section 5.5.2) of the field depending on scientific trends and paradigms that are dominant at any given time (section 5.5.3).

5.5.1. The Narrow Physico-processual Conception of Information Science

Definitions given in the early decades of the existence of IS, roughly between the 1970s and the 1990s, bore traces of Shannon's mathematical theory of information. In these first definitions, information was perceived as a physical entity which can be processed in a linear fashion from one stage to another of its life cycle. Harold Borko's widely cited definition of IS is a prime example of this processual conception of information:

> Information science is that discipline that investigates the properties and behaviour of information, the forces governing the flow of information and the means of processing information for optimum accessibility and usability. It is concerned with that body of knowledge relating to the origination, collection, organization, storage, retrieval, interpretation, transmission, transformation, and utilization of information. (Borko, 1968)

Borko mentions as his source of inspiration three earlier definitions of IS given by Robert S. Taylor in the *Annual Review of Information Science and Technology* (ARIST) in 1966. It is however noteworthy that at the time he gave this definition, Borko was working for "Information Systems Technology, Systems Develop Corporation", a software company based in Santa Monica. A psychologist by training who worked for the US Army

[26] Rojas (1983).

and for Rand corporation[27], Borko's conception of IS would have been influenced by the processual and behavioural perspective preeminent in psychology. His definition of IS is widely cited and adapted in different linguistic versions. Echoes of it have permeated the definitions given by different European scholars of IS in the 1970–1990s. Božo Težak, the Yugoslavian LIS pioneer gave a definition of informatology which echoed Borko's definition of IS:

> The very name of INFORMATOLOGY is meant to designate theory and practice of emission, transmission, accumulation, selection and absorption of information (the so-called e-t-ac-s-a complex[28]) and is wider and more exact than terms such as informatics, information sciences and services. (Težak, 1969a)

About the purpose of IS, Težak (1971) wrote that "Information science should investigate cognitive processes rather than simply technical, i.e. communication processes" (cited in Tudjman, 2010, p. 268).

The fact that Težak equated "technical" with communication once again reflects his chemical background. Communication is not seen by him as a relational phenomenon involving humans but as a process involving the linear transmission of data from one point to another. In line with the cybernetic theory of Norbert Wiener, Težak was persuaded that information and communication were two essential phenomena necessary for the advancement of society. He was quoted as saying: "It is unquestionable that the most essential changes in the development of the human species are closely connected with information and communication" (B. Težak, 1978 cited in Dj. Težak, 2010e, p. 301).

Težak developed the concept of "E-T-AC-S-A" as the conceptual basis from which to study information, communication and documentation (INDOC). In another article published in the first issue of *Informatologia Yugoslavica*, he described the INDOC concept in more details:

> All the structures that appeared under special forms and functions in the field of information, documentation and communication in the course of history, were but parts of a unique complex in which the components of creator and user, producer and consumer, as well as input and output of information are the most important. Just because of the intrinsic need to communicate, extrinsic emanation of transmission, accumulation and selection of communication and information

[27]Harold Borko's (1922–2012) biography is available at http://newsroom.ucla.edu/faculty-bulletin-board/in-memoriam–harold-borko–information-science-innovator.

[28]This is a central concept in Težak's understanding of information and communication processes to which we will return later.

> in all possible media are more and more required. That is why relations which will put into evidence the role of all necessarily integrating and connected factors, have to be permanently laid down by any analytic or synthetic information processing. Abstractly, there are at least five such factors: (1) emission, (2) transmission, (3) accumulation, (4) selection, and (5) absorption of information, consequently always the e-t-ac-s-a complex. Therefore in every phenomenon in the communication field, being a result of inside or outside processes and operations, one or more notions comprised in the e-t-ac-s-a complex are involved. (Težak, 1969b, pp. 1–2)

The emphasis placed by Težak's on the "e-t-ac-s-a" process reflects yet again his natural sciences background and his physico-processual conception of INDOC where the socio-cultural and historic dimensions are noticeably absent. His conception of INDOC as a closed system becomes even more clearer in a subsequent passage explaining the functioning of the "e-t-ac-s-a complex":

> The information, documentation and communication (INDOC) system is a complex of interactions where an element, datum, message, information, is transmitted, stored, kept, retrieved and used, mostly by means of a sort of material medium-document, so that the communication chain is a closed system from emission output, through transmission, accumulation and selection, to retrieval and input, i.e. absorption. (Težak, 1969b, p. 3)

Each of the components of the "e-t-ac-s-a" complex is seen as a subsystem in a complex that Težak represents as a tetrahedron, in which, according to him, one or two points (emission and absorption) "is changing permanently" (Težak, 1969b, p. 4) (See Figure 5.1 hereafter).

Težak's conception of IS and of the INDOC processes reflect the "scientism" current that permeated the social sciences in the 1960s, where in order to legitimise a field as a "science", its scholars sought to elaborate scientific theories or laws that governed the basic phenomenon they claimed to study.

Traces of this processual conception of information and of IS appeared in the writings of the first generation IS Professors in France. For instance, for Yves Le Coadic (1994, p. 28), it is concerned with "the study of general properties of information (nature, genesis, effect), of processes and systems of construction, of communication and of usage of that information[29]".

[29]In French in the original text: "*l'étude des propriétés générales de l'information (nature, genèse, effets), des processus et des systèmes de construction, de communication et d'usage de cette information*".

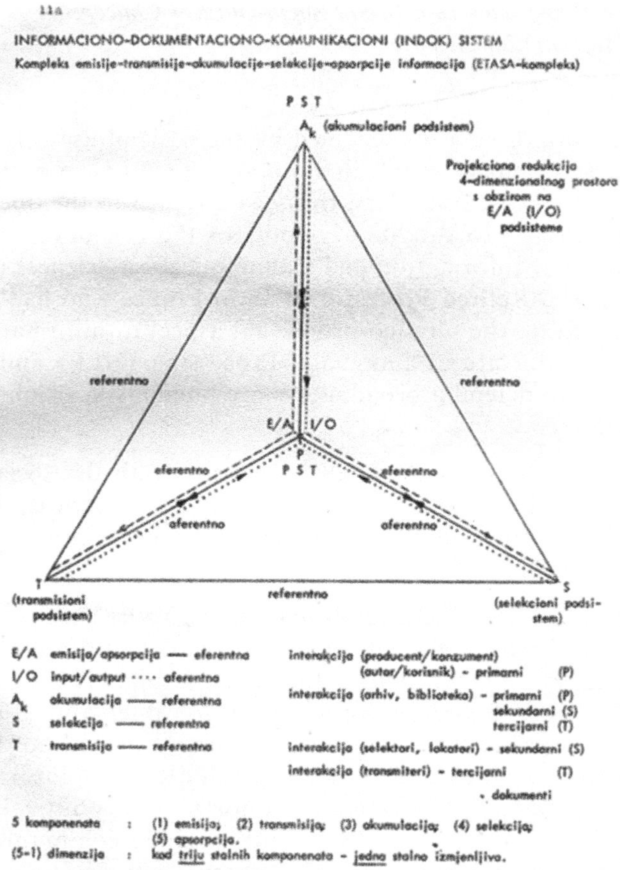

Figure 5.1: The E-T-AC-S-A Complex Figure. *Source*: Težak (1969b).

In France, the emergence of IS was separated from its library arm which remained the purview of a few special schools. Although the French pioneers were not from the sciences but from literatures, they were also preoccupied by building a science and a significant number of second generation scholars of ICS came from the sciences (physics, computer science, mathematics, statistics).

These first attempts to model IS after the engineering and formal sciences by adopting definitions that echoed Shannon's mathematical model of information or Wiener's cybernetics theory of communication have been repudiated by a second generation of IS scholars who have sought to bring the field back to its social, cultural and humanistic roots.

5.5.2. *The Swing back to a Broad Socio-cultural Conception of Information Science*

From the 1990s, there was a noticeable effort to cast IS as more of a social and cultural field preoccupied by the societal uses of information and the societal impact of information and communication technologies (ICTs). For French IS scholars, this was a return to the documentation origins of LIS and to the literary roots of the founding fathers of the interdiscipline of Information and Communication Sciences (ICS). Jean-Paul Metzger, a Retired Professor of IS in France who had based most of his research on the physico-processual aspect of information by having recourse to Natural Language Processing (NLP) and Computer Science came to defend a broad and more humanistic vision of IS at the end of his career:

> information science is essentially concerned with the social elaboration and with knowledge sharing. All types of knowledge is concerned, be it practical, technical, scientific, encyclopeadic or other. (Metzger, 2002, p. 19)

This trend to return LIS to its humanistic roots is equally visible in North America where there have been many calls for a more humanistic and socially engaged field rather than an engineering one. In his book *Information and Information Systems*, Buckland (1991) argued that data processing (information technology) should not be included in information studies and that an "exploration of information systems" must "include the social, economic and political contexts". He precisely advocated that IS should "include communication both at interpersonal and mass levels". Information studies without this social dimension would, according to him, be incomplete (Buckland 1991, p. 9).

Marcia Bates (1999, p. 1048), the "*Grande Dame*" of IS in North America also stated that the field is concerned with "human-produced information, and therefore, how human beings relate to this information – how they seek it, use it, ignore it, retrieve it – is of central research importance also. [...] in information science we study that process in service of information transfer".

For many second-generation IS scholars (Bates, 2007; Buckland, 2010), societal need is what gives IS its *raison d'être*[30] because information in itself is not important. What is more important is its relation to knowledge and to communication (Buckland, 1999b).

---

[30]Reason for being.

Tom Wilson (2010[31]), the Emeritus Professor of the University of Sheffield also exhorted IS scholars worldwide to adopt a broad social and humanistic conception of the field in his speech given at the occasion of being awarded the Doctor Honoris Causa of the University of Murcia, September 2010:

> So to conclude, let us not restrict ourselves to grubbing around in the garden patch of a limited, little information science, restricted to the relationship between information and machine. Instead, let us expand, reach out, embrace and explore the wider world of information to develop a vision of information science as a central, synthesising discipline in understanding not simply information, but the world we live in. Because the world we live in is surely a world of information.

Arguing for the specificity of the French IS against a perceived "anglophone one", Hubert Fondin, a retired French IS scholar held, unbeknownst to him, similar views with anglophone IS scholars about the necessity for IS to be part of the social sciences and humanities (Ibekwe-SanJuan, 2012a). Like Buckland and Bates did before him, Fondin argued that the French IS was concerned with the two arms of ICS, communication and information because the goal of every information processing, be it with technological tools, is to interact and communicate information to humans:

> Information science is one of the humanities and social sciences. Empirical phenomena that can be observed in information seeking (retrieval) can be seen as facets of a participation to a structure of communication, constructed by actors, with its instituted framework and its rules of exchange, in order to face a particular situation. There are naturally information-documents, tools and techniques, but those are only means, elements in a role play.[32] (Fondin, 2002, p. 115)

[31]Published in *Information Research*, 15(4), paper 439. Retrieved from http://informationr.net/ir/15-4/paper439.html.

[32]Our translation of his original text in French: "*La Science de l'information est une des sciences de l'homme et de la société. Les phénomènes empiriques que l'on peut observer dans la recherche d'information sont ainsi vus comme les facettes d'une participation à une structure de communication, construite par des acteurs, avec son cadre institué et ses règles d'échange, pour faire face à une situation particulière. Il existe naturellement des documents-informations, des outils et des techniques, mais ce ne sont que des moyens, des éléments dans un jeu de rôle*" (Fondin, 2002, p. 115).

Coincidentally, Tom Wilson deplored the lack of studies on both information and communication sciences whereas the two disciplines shared obvious links:

> It is also evident that the link between information and communication is necessary and strong: and yet, there is little research connection between communication studies and information science.[33]

One would have expected that having merged the two disciplines into one interdiscipline, the French ICS would set an example in investigating the links between both concepts. However, the ICS literature shows the co-existence of separate research communities, each with its own research traditions and publication channels (LIS, communication studies, media studies, cultural studies, journalism). Hence, the ICS is more a case of cohabitation than a real melding of two or more fields.

5.5.3. *The Oscillation between a Broad and Narrow Focus of LIS*

The history of LIS has been marked by this dance between narrow and broad conceptions, as technology evolved and its focus shifted from bibliography to documents and to information. As Ørom (2000, p. 15) observed:

> Before the Second World War, library science and documentation viewed its object in a historical and social context. After the war information science emerged, though it was first institutionalised in the sixties. The history of information science in its first decades is characterised by a double movement towards a broader scope and a narrower focus. This movement is most marked at the transition from library to information science.
>
> On the one hand the scope of the discipline was widened and generalised beyond the context of traditional library and information institutions, on the other hand the focus became narrower. In the thirties, library science had a quite coherent frame of understanding in the history of civilisation and in the notion of social functions. In the fifties, information retrieval research developed a scientific foundation. To put it briefly, the broadening of the scope and the narrowing of the focus also meant a transformation of the theoretical foundation from a humanistic and social science context to a scientific basis. In a way we could say that we had two unrelated corpora of empirical knowledge.

---

[33]Original text accessible at http://informationr.net/ir/15-4/paper439.html.

Bates (2005) attributed this swing to scientific trends that favour certain theories and paradigms at a given time while decrying the others:

> In more recent years, there has been a reaction to this approach [extreme scientism and logical positivism], with a concomitant swing towards the use of what are essentially humanities methods in the social sciences. Now the fashion is to deride the very scientific techniques so recently valorized and to insist that only highly qualitative and subjectivist methods produce credible results. Hermeneutic interpretation, detailed participant observation and historical analysis, among others, are now the methods of choice. Nowadays, it is seldom remembered, however, that the logical positivist approach was itself a reaction to what were deemed ineffective subjectivist research philosophies that preceded it. [...] In like fashion, attitudes toward information itself have swung between highly objectivist and subjectivist interpretations.

This swing between a broad and a narrow focus has marked the development of LIS in European countries. From the mid-seventies till the present, three approaches to LIS co-existed in the Nordic countries we studied, embodied by the scholars who promoted them: the information technology-computer science (IT-CS) approach which reflected the narrow conception of LIS; the sociological and domain analytic approach; and the literary-bibliographic-cultural-historical approaches which both reflected broad conceptions of LIS.

The broad conception is mainly motivated by the desire to study problems from different perspectives while the narrow focus is mostly driven by the desire to build a science and a discipline. In contrast to Finland where the development of LIS was driven by the desire to build a discipline in the natural sciences sense of the word, the other tree Nordic countries favoured a broader conception of the discipline:

> Since their transition from vocational training to academic education in the early 1990's, the major Nordic schools in Denmark, Norway and Sweden opted for keeping a broad approach to library and information science, while Finland, where the discipline of library and information science was instituted almost a generation earlier, chose the consistent disciplinary approach, signifying information science and its core concepts as the basis and framework for research and education. According to a recent study of the history of Finnish information science this has brought about a separation between the discipline of information science and the profession of librarianship in Finland (Mäkinen et al., 2016). (Limberg, 2017, p. 4)

Limberg (2017) observed that the Swedish LIS distanced itself from the prevailing narrow focus on information and information processes and instead cast LIS as a multidisciplinary field integrating not only information but also culture, concerned not only with information processes (the analytical side) but with mediation, involving not only libraries but also other types of institutions concerned with the mediation of information. This is reflected in the definition of LIS given in a 1991 call from the Swedish Council for Planning and Co-ordination of Research for the first Swedish chair for a Professor of library and information science at the University of Gothenburg (occupied by Lars Höglund in 1993):

> The discipline takes its point of departure in problems related to the mediation of information or culture, stored in some form of document. The objects of study are processes such as information provision or the mediation of culture, as well as libraries and other institutions with similar functions, involved in this process. The discipline has connections to a range of other disciplines within the social sciences, the humanities and technologies. (FRN, 1989, p. 85, my translation[34] cited in Limberg, 2017)

The reason for the broad Swedish conception of LIS was conjectural. The first professors in LIS came from the humanities and social sciences, and so had broader interests. They were not specifically inhabited by a desire to make LIS a science:

> To sum up, the combination of a broad definition of the discipline, and the attitudes and interests of the leaders who were to direct the implementation of education on all levels, including PhD, led to an open approach to library and information science focusing on capturing and analysing what seemed to be interesting problems to be investigated, adopting various theoretical approaches from library and information science as well as from other disciplines, and thus building the discipline according to the explicit principle "you take what you've got"[35] as formulated by Romulo Enmark (1991). (Limberg, 2017, p. 4)

As Bates (1999) stated, LIS needs theories and methods from the sciences, the social sciences and the humanities to solve all the problems that it aims to investigate:

> A final comment on methodology: regarding the great methodological shift sweeping through the social sciences, the shift to

[34] "FRN, Swedish council for planning and co-ordination of research, was a government body for funding research". Cited in Limberg (2017, p. 2).

[35] "The saying comes from an 18th century famous Swedish cook, Cajsa Warg, to characterise how to pick ingredients for a recipe". (Cited in Limberg, 2017, p. 4).

> the qualitative, multiple-perspective, post-Modernist approaches – these new techniques simply add to and enrich the armamentarium of techniques available to the information scientist for studying the subject matter of our field. For reasons that have already been argued, this field requires multiple methodological approaches to conduct its research. In mid-20th century social science we have had a series of waves of methodological fashion – each wave declaring the prior approach to be hopelessly bankrupt and inadequate. It is to be hoped that it is finally recognized that all of these methodological approaches can be powerful and useful – especially in information science. (Bates, 1999, p. 1049)

Ørom (2000, p. 23) also concluded that the changing paradigms and conceptions of LIS are the results of "historical changes of the universe of knowledge and needs for information". In the pre-war era (Second World War), libraries were viewed in the context of history of civilisation as social institutions. With the emergence of information technology in the 1960s came the physical paradigm which entailed a shift from libraries to information. On the heels of the physical paradigm came the cognitive paradigm which sought to give IS a theoretical grounding for its research on the use of online search systems. Lastly, a socio-historic paradigm emerged in the 1990s which tried to re-integrate the historical and social contexts that had been lost or forgotten during the physical and cognitive turns.

5.5.4. In Search of the "Object" of a Discipline: The Circle of Relativism

Disciplines have two ways of justifying their existence: ontologically, by what they say they are or what their "object" or subject matter is; or secondly, epistemologically, through the theories, concepts and methods by which they arrive at knowledge they claim to elaborate. Attempts to legitimise IS have tended to favour the ontological viewpoint because it seems easier to identify a unique "object" and lay claim to it. Hence, LIS could claim to be about the study of bibliography, of documents, or of information, according to the shifts in "disciplinary objects" across time. For instance, Jean Meyriat claimed that IS was distinctive from other disciplines because although information was an "object" shared by the whole of humanity and by all the other disciplines, it was the central paradigm of no other discipline because no other discipline was dedicated to "explaining the nature of this phenomenon, to establishing its working conditions, to identifying the theoretical knowledge upon which technicosystems that are meant to mobilise and manage information can be built" (Meyriat, 1986).

The ontological justification for the existence of a discipline is however problematic. IS does not have exclusive rights to the study of information. It is also the central paradigm of many other fields like philosophy, education, history, journalism, communication, computer science, physics and biology. Many have since realised that this ontological path is not fruitful:

> Debates over the "proper" definition of information science, as of any field, are fruitless, and in expectations naive. Information science, as a science and as a profession, is defined by the problems it has addressed and the methods it has used for their solutions over time. (Saracevic, 1999)

Yves Jeanneret and Bruno Ollivier, two ICS scholars in France illustrated the theoretical impossibility of delineating the frontiers of any scientific discipline by choosing an "object" and claiming exclusive rights to it:

> Disciplines have more or less an object, meaning that this object can be more or less clearly defined but even for the well established disciplines, it is not certain that this object does not slip away. If demography deals with variations in populations, it can be said to have a recognisable object, just as gastroenterology or astrophysics do (at least for the novice). The object of sociology is already more difficult to circumscribe, it can be everywhere. That of linguistics is not at all clear for the non-linguist since it is based on a class created by linguists who invented two categories – language and speech, and decided to study what the former meant to them. As for philosophy, its object is most in(de)finite.[36] (Jeanneret and Ollivier, 2004, p. 14).

[36]My translation of the author's original text: «*Les disciplines ont plus ou moins un objet. Entendons par là que celui-ci se définit plus ou moins aisément et que, même pour les plus assurés, il n'est jamais certain qu'il ne se dérobe pas. Dans la mesure où la démographie traite des variations de population, elle a un objet saisissable, comme la gastroentérologie ou l'astrophysique (au moins pour le non spécialiste). L'objet de la sociologie est déjà moins facile à circonscrire. Il peut être partout. Celui de la linguistique ne l'est pas du tout pour le non linguiste, puisqu'il repose sur une catégorie produite par les linguistes qui ont inventé deux catégories, langue et parole, et décidé de traiter de ce que recouvre pour eux la première. Quant à la philosophie, son objet est le plus in(dé)fini* (Jeanneret & Ollivier, 2004, p. 14).

Therefore, the justification and definition of a discipline cannot be fruitfully based on a specific "object" because "objects" do not belong to disciplines, especially information and communication which are fundamental phenomena of the universe and are everywhere.

This point was well made by the twentieth-century philosopher Karl Popper (1963):

> Disciplines are distinguished partly for historical reasons and reasons of administrative convenience (such as the organisation of teaching and appointments), and partly because the theories which we construct to solve our problems have a tendency to grow into unified systems. But all this classification and distinction is a comparatively unimportant and superficial affair. We are not students of some subject matter, but students of problems. Any problem may cut right across the border of any subject matter or discipline.[37]

There now seems to be a consensus amongst scholars from several disciplines that disciplines are defined not by their "objects" (or rather subjects) but by their "project" (see Ibekwe-SanJuan, 2012a, for a discussion). This naturally leads to considering the alternative justification, the epistemological one, i.e., the manner in which this subject is studied and for what purposes and how that distinguishes a field from the others.

Using a forceful historic analogy, Ørom (2000, p. 24) considered that a major turning point in the history of conceptions of LIS was "the transition from the Ptolemaic picture of the information universe to the Copernican. The Ptolemaic picture is characterised by the central and well-defined position of the library in the information universe; furthermore, it is characterised by strong cultural traditions and an expanding, but manageable and conceptually rather stable corpus of scientific, scholarly and practical knowledge. In the Copernican universe of information, the libraries and information institutions are but planets, knowledge is becoming fragmented, depersonalised, scattered among many messages and subject to rapid conceptual change. The needs for information are diversified, determined by social change. This new universe of information has been a challenge to Information Science since the sixties."

LIS continues to reflect this fragmented and scattered knowledge landscape where many paradigms and theories co-exist side by side more or less harmoniously. It continues to encompass a large spectrum of research topics ranging from biblio-sciento-informetrics, information seeking behaviour, information retrieval systems, information literacy, knowledge

---

[37]Popper (1963).

organisation to information architecture and data privacy as well as theoretical and historical research. The broad conception of LIS which the humanities and social sciences pioneers sought to promote appear to have prevailed but the quest for internal epistemological coherence remains, like Sisyphus's task, unattainable.

Conclusion

This book attempted a history of the emergence of LIS in seven European countries with the aim to identify their common and distinctive features and more importantly to uncover the network of influences and the circulation of ideas amongst European pioneers.

In terms of shared features, three traits appeared to run through all the countries we studied: the evolution of the earliest LIS institutions from vocational to academic structure; the multidisciplinarity of the teaching staff and the network of influences which enabled the cross-fertilisation of ideas.

The first shared trait was the fact that the earliest formal training programmes in all the countries were rooted in librarianship or "library economy" as it was called then which was not considered a science but a technical skill that did not warrant academic education. From the 1970s, as library schools began collaborating with universities, the latter slowly absorbed and transformed library education, broadened its teaching staff base to include academics. In the first decades of university existence, teaching staff were recruited for their professional expertise in one or more aspects of library service and happened to have diverse academic origins. The majority of the teaching staff, if they had a prior academic background, came mostly from the humanities and did not have a PhD. As LIS gained acceptance as an academic field in 1980-90s, it gradually became a requirement for its Faculty to have a PhD and to carry out research. LIS thus witnessed a reversal of the trends in that instructors are now mostly academics without prior professional experience in librarianship, in documentation and in the other information professions. This has caused hostility and in some countries like France, a split between the information professionals on the ground and the academics in universities who are in charge of training future information professionals.

A second common trait of LIS worldwide is that LIS has always been a multidisciplinary discipline from birth. This resulted in the co-existence of different research traditions and paradigms in IS which in turn resulted in a divide between the subjectivist paradigm and the physico-objectivist paradigm, with little or no interaction between the two.

Scholars from the humanities and social sciences who were less inhabited by the desire to build a science upheld a broader conception of the discipline in which the three subfields, information science, documentation and librarianship, cohabited, thus they upheld an LIS with the "L". Scholars in this research tradition championed the subjectivist paradigm and developed the more historical and theoretical research on information literacy and bibliography using qualitative methods.

On the other hand, scholars who were more preoccupied with building a discipline and a science (in the more restrictive English sense) tended to focus solely on the need for automation of library and documentation procedures (automatic indexing and searching of online bibliographical records). Their quest for laws governing the production, the processing and dissemination of scientific and technical information (STI) resulted in a narrow conception of the field which implied a preference for the physico-objectivist paradigm and a reliance on theories and methods from the sciences (mathematics, physics, statistics, computer science). This led to the development of research in information systems, information retrieval, user studies (quantitative and cognitive aspects) and the metrics (biblio/sciento/info/webometrics) within IS.

Thirdly, our cross-national study showed that European pioneers of LIS were well integrated into the network of international associations such as the UNESCO, FID, ASLIB, IFLA, ASIST. Our track for the paternity of the two umbrella terms "informatology" and "informatics" illustrated this network of influences between European pioneers as they borrowed concepts and ideas from one another and adapted them to their national context, sometimes with similar meanings, sometimes with varying or even opposing meanings. This network of influence is more noticeable in the Nordic countries where there has always been a tradition of cooperation, competition and mobility in several sectors, including higher education and research which promoted a spirit of emulation and innovation in the development of LIS.

A recurrent concern in all the countries we studied is LIS's institutional fragility. There is a sense of general decline in the number of professors and students enrolled in LIS institutions owing to organisational, structural and economic adjustments within the higher education institutions. This has resulted, in certain countries, in a loss of autonomy of LIS departments and their absorption into other departments and faculties. There is a heightened sense that LIS is facing increasing challenges from neighbouring social sciences and humanities disciplines amongst whom the most cited are communication, journalism, media studies, culture, sociology, etc. This sense of anxiety is more acute in France and in Portugal where the discipline has never known a separate existence but was integrated into a wider humanities field right from birth notably, communication and media studies. A recent bibliometric study by Ortiz-Repiso Jimenez and Pacios (2018, p. 154) on the state of European LIS education from 2000–2017, i.e. since the creation of the European Higher Education Area (EHEA) and after the Bologna Process, confirmed this downward trend: student intake is declining in western countries (Europe and the United States) while other regions are witnessing a boom in LIS education. The middle east is witnessing an upward trend in LIS intake of students due to a concerted effort to promote the field.

In terms of distinctive features, the European countries studied displayed differences in their models of geographic or "territorial expansion" and in the external influences that shaped the way LIS was conceptualised by their pioneers.

Concerning their territorial expansion, two opposing models were observed: a centripetal and a centrifugal one.

The centripetal model was displayed in two countries: Denmark and to a lesser extent France. In Denmark, Copenhagen was and still is the epicentre of LIS development, even if a branch of the Royal School of Librarianship (RSL) was opened in Aalborg in 1973, it remained a secondary one offering training only for public libraries. France is a very centralised country where the seat of power is concentrated in Parisian institutions. Between 1821 and the 1970s, LIS institutions were concentrated in Paris. This situation has changed since and there are now several centres of LIS education in France

Sweden, Norway, Spain and Portugal on the other hand, displayed a centrifugal model of territorial expansion.

The case of Yugoslavia is rather peculiar. In its first decades of existence, territorial expansion of LIS displayed a combination of a centripetal model with a spiralist or cybernetics twist. The first successful LIS institutions were all located within the University of Zagreb and nested within one another, like Russian *matryoshka* (nesting) dolls.

The second distinctive feature we observed concern the external influences that shaped the way LIS was conceptualised in each country. Unsurprisingly, the external influences reflected each country's historic ties with the two former imperial powers: the Anglo-American and French spheres of influence.

The Nordic countries and Yugoslavia were predominantly under Anglo-American influence, while LIS development in Latin Europe was clearly modelled after the Francophone (Belgium and France) documentation tradition.

Epilogue

Much remains to be done in order to connect the histories rendered here with other national histories written in languages other than English. Germany's LIS history is yet to be brought to an international audience. The situation of LIS in Austria shares some characteristic traits with the European countries we studied but also differs significantly in that LIS higher education is quite a recent phenomenon in this country. Like in the other European countries we studied, the first library education was established in the late 1920s as an "in-house training" offered by the Austrian National Library under the supervision of the Ministry of Science. The education of academic librarians was separate from that of librarians that worked in public and school libraries. It was only in 2004 that a Master of Science degree in LIS was established and run jointly by the University of Vienna and the Austrian National Library. There is as of now no cohesive body of university programmes in LIS connecting the undergraduate to graduate and post-graduate levels (Hörzer and Schögl, 2018). A more in-depth study is needed to identify Austrian pioneering figures as well as their possible links with other European and international pioneers. Further towards the East, Poland and Russia's LIS histories are still to be brought to an international audience. Coincidentally, 2018 marks the 100th anniversary of LIS education in Denmark and in Russia since the first "Library Institute" was founded in 1918 in Saint Petersburg by Nadezhda Krupskaya, the wife of the Russian revolutionary leader Vladimir Lenine (Brezhneva, Gruzova, Krymskaya and Soloveva, 2018).

Further afield, the histories of LIS in Asian, African, Australia and Latin American countries also need to be connected to that of European and North American countries. Concerning Australia, Howard (2018) observed that despite "a decline in universities offering LIS education and the comparatively low student numbers, Australia still has more LIS programmes than countries such as the UK, USA and Canada". She suggested that the size of the country may explain why many LIS programmes were made available to cover the sheer expanse of the territory but that was before information and communication technology enabled online learning which has made the multiplication of LIS programmes unnecessary. Australian LIS programmes have to balance two contradictory constraints.

On the one hand, in order to obtain accreditation from the Australian Library and Information Association (ALIA), they must be library-focused because of the expectation from this organisation that LIS degree holders are destined to work in libraries. On other hand, LIS scholars are aware of the necessity to create *niche* programmes that respond to current societal needs in information systems. However, with very little faculty resources, this equation is difficult to balance. This has led to a gap between the names of LIS university programmes which are called "Information Management, Information Studies, Information and Library Studies, and even Business Information Systems" and their real content (Howard 2018: 106). The tensions caused by the geographic, economic, political and institutional factors on LIS education and research are not peculiar to Australia. They need to be cast in a more international setting to determine how they have been shaping the way LIS education and research are developing and evolving worldwide.

We hope that other scholars will take up the mantle of studying how LIS developed in different regions of the world not in a "siloed national histories" perspective but as comparative histories which will help uncover common and distinctive features that will shine a light on the past, connect it to the present and illuminate the future paths of Library and Information Science worldwide.

Appendix

A1. Interviewees CROATIA – 9 April 2017

Neva Tudor-Silovic

Neva Tudor-Silovic holds a graduate degree in English and Sociology. She worked in Germany and Denmark for a long before returning to Croatia. She was a close collaborator of Božo Težak and worked as his assistant in the Referral Center from 1974 to 1994. She was in charge of international cooperation and organising conferences in Dubrovnik on Scientific and Technical Information.

Tatjana Aparac-Jelusic, Emerita Professor of LIS

Tatjana Aparac-Jelusic was head of two LIS departments in Zadar and Osijek. She started both programmes in Library Science which then evolved to Information Sciences. Now the departments cover Library Archives Museums (LAM).

Although "officially retired", Tatjana continues to be involved in several LIS organisations and projects. She is currently President of the Croatian Council on Libraries which produced the strategy/action plan of Croatian librarianship till 2021.

A2. Interviewee for Denmark, Sweden, Norway, 4–7 May 2017

Niels Lund Windfield was born in 1949 in Copenhagen, Denmark. He trained as a historian and ethnologist. From 1975 till 1988 he was Associate Professor at the Royal School of Library and Information Science in Denmark. He went on to become the first employee and Full Professor at the Department of Documentation Studies, University of Tromsø, Norway, in 1996. He was responsible for the initial development of this programme at both graduate and graduate levels. Lund has twice been Visiting Professor at the University of California, Berkeley (2001/2005–06) and Visiting Scholar at CCRMA, Stanford University in 2011. He retired in 2014 as Professor Emeritus.

A3. Interviewees for Spain, 29–31 May 2017

The Barcelona Group

(1) **Ernest Abadal** majored in Philosophy before obtaining a diploma in LIS in 1987. He worked as a teacher at the Autonoma University of Barcelona in the Faculty of Communication Science where he taught a subject called "Documentation applied to media" for 10 years. He joined the Faculty of Documentation and Librarianship in 1992 and taught information technology, then methodology and information and society. He was the Dean of the Faculty of Biblioteconomia i Documentació, Universitat de Barcelona from 2012 to 2018.

(2) **Tomás Baiget** is a retired but still very active consultant in IS. He is the Director of the scientific journal *El Profesional de la información*. He first worked in the telecommunications industry with Philips before being given an opportunity to work on databases in Europe at the European Space Agency (ESA) in 1973. He began his teaching career on online searching in Madrid where he lived. He was invited to the University of Barcelona around 1985/6 as a consultant. He worked as a Project Head at a consortium of information integrated by local authorities (City Council, Province Council, Metropolitan Area and Chamber of Commerce), converted today into the Catalan Statistics Institute (Idescat). He helped private and public sectors to begin online searching when this subject was not taught in universities. He was member of the Organizing Committee of online meetings in London. He is a creator and manager of https://directorioexit.info with 4.000 records of librarians and information specialists. He specialised on scientific editing and publishing. He gives talks in two to three universities per year in Spain and Latin America on how to publish papers and on how to edit scholarly journals as well as webinars organised by Sedic. He has written *Historia de Fesabid* (http://eprints.rclis.org/3110) and has just initiated the project *"Cronologia de la Documentación Española"* (*CroDoc*) accessible online at http://crodoc.es.

(3) **Agustí Canals** is a physicist with an MBA and a PhD in economics. He created the new Documentation degree at the Open University of Catalunya, a brand new programme that is completely online. At the

time, there was no Bachelor's degree in LIS. The idea was to create something that will have a broad perspective unlike the traditional LIS programmes focused on producing librarians and archivists. Hence, content related to Information Management, Competitive Intelligence and InfoVisualisation were introduced in the new programme.

The Madrid Group

(1) **Mercedes Caridad Sébastian** holds a PhD in Journalism. Her career in LIS began in 1978 as a Professor at the Faculty of Information Science at Complutense University, Madrid, where she is currently working. She worked with Lopes Yepes and Felix Sagredo and Desantes. She also worked in the information industry on databases (Dialog, etc.). She was Head of the Department of Library Science, University Carlos III, Madrid.

(2) **Virginia Ortiz-Repiso Jimenez** has been a Professor since 1996 at the Library and Information Sciences Department, Carlos III University, Madrid. She has been involved in Bologna Process at the European Higher Education Area (EHEA) at national and international levels. Her main research and teaching interests are in managing technology in digital libraries, digital collections, searching and retrieving information in electronic environments, digital humanities, and Open Access movement. Her works are published in national and international journals. Since March 2015, she is the iDean representative of the iSchool Universidad Carlos III, Madrid.

(3) **José Antonio Moreiro González** holds a PhD in History on the figure of Agustin Millares Carlo, a pioneering figure of Spanish LIS.

Previously, he had taught at the UNED (1982–1986), at the Universidad Complutense (1986–1989), the Universidad de Murcia (1989–1991). Since 1991, José Antonio Moreiro is a professor of Library and Information Science department at Universidad Carlos III de Madrid. His teaching and research focus on knowledge organization, semantic vocabularies, information science theory and labour market in information science.

He held various administrative positions at the Carlos III University of Madrid. He was the Head of the Department of Library and Information Science (1992–2000), Dean of the Humanities, Communication and Documentation Faculty (2002–2010), and University ombudsman (2012–2018). He has been a Visiting Professor in 40 universities abroad.

(4) **Maria-Antonia García Moreno** began working in the LIS field at Alcala de Henares University in Madrid in 1984 as a documentalist. She then

worked in a documentation centre at the EU (1986–95). She worked as a professional in private companies in different sectors. In 1992, she began teaching at Carlos Madrid III in its Faculty of Social Science (within which LIS training was offered), and then from 1995 at the Complutense University at the Faculty of Documentation.

(5) **Fernando Ramos** majored in law and journalism at Complutense University in Madrid. He has been an Associate Professor at Complutense University while working as a professional in the Association of Electricity Companies. He was the Dean of the Faculty of Documentación in this University between 2005 and 2011.

A4. Interviewee for Portugal

Fernanda Ribeiro is Full Professor at the Faculty of Arts and Humanities of University of Porto (Portugal) where she began her academic career in 1989. Previously she worked as librarian-archivist in the Municipal Archives of Porto City Council. She has been Director of the Bachelor degree programme in Information Science (2003–2014) and Director of the Doctoral Programme in Information and Communication in Digital Platforms (2013–2014). She was Scientific Coordinator of the research centre CETAC.MEDIA (Centre of Studies in Technologies and Communication Sciences) from 2012 to 2014. Since October 2014, she is Dean and Head of the Scientific Board of the Faculty of Arts and Humanities of University of Porto.

References

General

Bates, M. J. (1999). The invisible substrate of information science. *Journal of the American Society for Information Science*, *50*(12), 1043–1050.

Bates, M. J. (2005). Information and knowledge: An evolutionary framework for information science. *Information Science Research*, *10*(4). Retrieved from http://InformationR.net/ir/10-4/paper239.html. Accessed on October 10, 2010.

Bates, M. J. (2007). Defining the information disciplines in Encyclopedia development. *Information Science Research*, *12*(4). Retrieved from http://informationr.net/ir/12-4/colis/colis29.html. Accessed on October 10, 2010.

Bawden, D., & Robinson, L. (2012). *Introduction to information science*. London: Facet Publishing.

Bobinski, G. (Eds.). (1986). Current and future trends in library and information science education. *Library Trends*, *34*(4), 537–559.

Borko, H. (1968). Information science. What is it? *American Documentation*, *19*(1), 3–5.

Buckland, M. K. (1991). *Information and information systems*. Santa Barbara, CA: Praeger.

Buckland, M. K. (1996). Documentation, information science, and library science in the U.S.A. *Information Processing & Management*, *32*(1), 63–76.

Buckland, M. K. (1999a). Overview of the history of science information systems. In M. E. Bowden, T. B. Hahn, & R. V. Williams (Eds.), *Proceedings of the 1998 Conference on the History and Heritage of Science Information Systems* (pp. 3–7). Medford, NJ: Information Today.

Buckland, M. K. (1999b). The landscape of information science: The American society for information science at 62. *Journal of the American Society for Information Science*, *50*(11), 970–974.

Buckland, M. K. (2007). On the cultural and intellectual context of European documentation in the early twentieth century. In B. R. Rayward (Ed.), *European modernism and the information society: Informing the present, understanding the past* (pp. 44–57). Aldershot: Ashgate. Retrieved from: https://escholarship.org/uc/item/2tb100vc

Buckland, M. K. (2012). What kind of science *can* information science be? *Journal of the American Society of Information Science*, *63*(1), 2012, 1–7.

Buckland, M. K. (2017c). Library technology in the next 20 years. *Library Hi Tech*, *35*(1), 5–10.

Buckland, M. K., & Liu, Z. (1998). History of information science. In T. B. Hahn & M. Buckland (Eds.), *Historical studies in information science* (pp. 272–295). Medford, NJ: Information Today. An updated version is available at http://escholarship.org/uc/item/2pp2781h

Buckland, M. K., & Lund, N. W. (2013). Boyd Rayward, documentation, and information science. *Library Trends*, *62*, 42–50.

Capurro, R., & Hjørland, B. (2003). The concept of information. *Annual Review of Information Science and Technology*, *37*, 343–411.

Cornelius, I. (2002). Theorizing information for information science. *Annual Review of Information Science and Technology*, *36*, 393–425.

Cronin, B. (1992). Information science in the international arena: An educator's perspective. *Aslib Proceedings*, *44*, 195–202.

Engelbart, D. C. (1963). A conceptual framework for the augmentation of man's intellect. In *Vistas in information handling* (Vol. 1, pp. 1–29). Washington, DC: Spartan Books.

Fasick, A. (1986). Library and information science students in current and future trends in library and information science education. *Library Trends*, *34*(4), 607–621.

Furner, J. (2010). Philosophy and information studies. *Annual Review of Information Science & Technology*, *44*, 159–200.

Hauk, K., & Stock, W. G. (2012). Pioneers of information science in Europe: The oeuvre of Norbert Heinrichs. In T. Carbo & T. B. Hahn (Eds.), *International perspectives on the history of information science and technology. Proceedings of the ASIS&T 2012 Pre-Conference on the History of ASIS&T and Information Science and Technology* (pp. 157–162). Medford, NJ.

Hjørland, B. (1998). Theory and metatheory of information science: A new interpretation. *Journal of Documentation*, *54*(5), 606–621.

Hjørland, B. (2000). Documents, memory institutions and information science. *Journal of Documentation*, *56*(1), 27–41.

Hjørland, B. (2017). *Library and information science*. In Maja Žumer (Eds.), ISKO encyclopedia of knowledge organization. Retrieved from http://www.isko.org/cyclo/lis. Accessed on December 13, 2017.

Hjørland, B., & Albrechtsen, H. (1995). Toward a new horizon in information science: Domain-analysis. *Journal of the American Society for Information Science and Technology*, *46*(6), 400–425.

Machlup, F., & Mansfield, U. (1983). Semantic quirks in studies of information. In F. Machlup & U. Mansfield (Eds.), *Studies of information: Interdisciplinary messages* (pp. 641–671). New York, NY: John Wiley & Sons.

Mikhailov, A. I., & Giljarevskij, R. S. (1971). An introductory course on informatics/documentation. Working document submitted by the FID to the UNESCO, 31 Aug. 1970, 208 pages. Retrieved from http://unesdoc.unesco.org/images/0000/000006/000676eb.pdf

Otten, K., & Debons, A. (1970). Towards a metascience of information: Informatology. *JASIS*, *21*(1), 89–94.

Popper, K. R. (1963). *Conjectures and refutations: The growth of scientific knowledge*. (p. 88). New York, NY: Routledge & Kegan Paul.

Rayward, W. B. (1983). Library and information sciences. In Machlup & Mansfield (Eds.), *The study of information* (pp. 343–363). New York, NY: John Wiley & Sons.

Rayward, W. B. (1994). Visions of Xanadu: Paul Otlet (1868–1944) and hypertext. *JASIS*, *45*, 235–250.

Rayward, W. B. (1996). The history and historiography of information science: Some reflections. *Information Processing & Management*, *32*(1), 3–17.

Rayward, W. B. (1997). The origins of information science and the International Institute of Bibliography/International Federation for Information and Documentation (FID). *Journal of the American Society for Information Science*, *48*(4), 289–300.

Rayward, W. B. (2003). Knowledge organisation and a new world polity: The rise and fall and rise of the ideas of Paul Otlet. *Transnational Associations*, *1–2*, 4.

Rayward, W. B. (2004). When and why is a pioneer: History and heritage in library and information science. *Library Trends*, *52*(4), 671–682.

Rojas, O. (1983). La investigación y el desarrollo integral de la Bibliotecología en la región. *Revista Interamericana de Bibliotecología*, *6*(1–2), 5–17.

Saracevic, T. (1999). Information science. *Journal of the American Society for Information Science*, *50*(12), 1051–1063.

Shera, J. H., & Cleveland, D. B. (1985). History and foundations of information science. *Annual Review of Information Science and Technology*, *12*, 249–275.

White, H. D., & McCain, K. W. (1998). An author co-citation analysis of information science, 1977–1995. *Journal of the American Society for Information Science*, *49*(4), 327–355.

Williams, R. V. (1997). The documentation and special libraries movements in the United States 1910–1960. *Journal of the American Society for Information Science*, *48*(9), 775–781.

Yang, S., & Feifeu, W. (2015). Visualizing information science: Author direct citation analysis in China and around the world. *Journal of Informetrics*, *9*(1), 208–225.

Zhao, D. Z., & Strotmann, A. (2008). Information science during the first decade of the web: An enriched Author co-citation analysis. *Journal of the American Society for Information Science and Technology*, *59*(6), 916–937.

France

Benoit, G. M. (2008). *Eugene Morel: Pioneer of public libraries in France*. Sacramento, CA: Litwin Books.

Bourdieu, P. (1976). Le champ scientifique. *Actes de la recherche en sciences sociales*, *2*(2–3), 88–104. doi:10.3406/arss.1976.3454

Boure, R. (Ed.). (2002). Quelle histoire pour les sciences de l'information et de la communication? In *Les origines des & Technology, Sciences de l'information et de la communication. Regards croisés* (pp. 17–44). Lille: Presses universitaires du Septentrion, Villeneuve d'Ascq.

Briet, S. (1951). *Qu'est-ce que la documentation?* Paris: EDIT.

Briet, S. (1953). Bibliothèques et centres de documentation technique aus Etats-Unis. Notes d'un voyage de quatre mois (Octobre 1951–Février 1952). *A.B.C.D* Archives, Bibliothèques, Collections, Documentation, 11, 299–308.

Buckland, M. K. (2005). A brief biography of Suzanne Renée Briet. In S. Cacaly, Y. Le Coadic (2007), Fifty years of scientific and technical information policy in France (1955–2005). *Journal of Information Science*, *33*(3), 377–384.

Buckland, M. K. (2017a). Some European contributions to information science. *Bulletin of the Association for Information Science and Technology*, *43*(3), 49–52. Retrieved from https://www.asist.org/files/bulletin/feb-17/Bulletin_FebMar2017.pdf

Buckland, M. K. (2017b). Before the antelope: Robert Pagès on documents. *Proceedings on the Document Academy*, *4*(2), 2-7. Retrieved from https://ideaexchange.uakron.edu/docam/vol4/iss2/6/

Buckland, M. K. (2018). Reflections on Suzanne Briet. In Widad Mustafa El Hadi (Eds.), *Fondements épistémologiques et théoriques de la science de l'information-documentation. Actes du 11e colloque ISKO France 2017* (pp. 10–21). London: ISTE Editions.

Buckland, M. K., & Liu, Z. (1995). History of information science. *Annual Review of Information Science and Technology*, *30*, 385–416.

Calenge, B. (2000). Formation et recherche bibliothéconomiques en France: lecture d'une complexité. *Bibliothek Forschung und Praxis*, *24*(1), 18–26.

Davallon, J. (2004). Objet concret, objet scientifique, objet de recherche. *Hermès*, *38*, 30–37.

Day, R. E. (2007). Suzanne Briet: An appreciation. *Bulletin of ASIST*, *33*(2), 21–22. Retrieved from http://www.asis.org/Bulletin/Dec-06/day.html

Day, R. E., & Martinet, L. (2006a). *Briet S: What is documentation?* English translation of the classic French text. Lanham, MD: Scarecrow Press. Retrieved from http://info.slis.indiana.edu/˜roday/what%20is%20documentation.pdf

Day, R. E., & Martinet, L. (2006b). Préface to Suzanne Briet's *Qu'est-ce que la documentation*. Retrieved from http://ella.slis.indiana.edu/˜roday/Briet_preface.pdf

Delmas, B. (1992). Une fonction nouvelle: Genèse et développement des centres de documentation. In M. Poulaine (Ed.), *Histoire des bibliothèques francaises* (Vol. 4, pp. 178–193). Paris: Promodis.

Delmas, B. (1993). L'INTD et son rôle dans la formation des documentalistes en France: 1932–1993. *Documentaliste - Sciences de l'information*, *30*(4–5), 218–226.

Ducolomb, L. (2012). Eugène Morel et la section des bibliothèques modernes. Une réflexion sur la formation des bibliothécaires au XXème siècle. *Bulletin de Bibliothèques de France (BBF)*, tome. 57(1), 35–39. Retrieved from http://bbf.enssib.fr/consulter/bbf-2012-01-0035-006

Fayet-Scribe, S. (1997). The cross-fertilization of the U.S. public library model and the French documentation model (IIB, French correspondent of FID) through the French professional associations between World War I and World War II. *Journal of the American Society for Information Science*, *48*(9), 782–793.

Fayet-Scribe, S. (2000). *Histoire de la documentation en France. Culture, science et technologie de l'information 1895–1937*. Paris: CNRS Éditions.

Fondin, H. (2002). La "science de l'information" et la documentation, ou les relations entre science et technique. *Documentaliste-Sciences de l'information*, *39*(3), 122–129.

Gardin, J.-C. (1964). *Le Syntol: étude d'un système général de documentation automatique*. Bruxelles: Presses académiques européennes.

Hassenforder, J. (1967). *Développement comparé des bibliothèques publiques en France, en Grande-Bretagne et aux États-Unis dans la seconde moitié du XIXe siècle (1850–1914)*. Paris, Cercle de la librairie.

Ibekwe-SanJuan, F. (2012a). The French conception of information science. Une exception française? *Journal of the American Society for Information Science & Technology*, *63*(9), 1693–1709.

Ibekwe-SanJuan, F. (2012b). Whither information science in France?: Special event at the ASIS&T 75th Anniversary Event on the History of ASIS&T and Information Science and Technology Worldwide, Annual Meeting of the American Society for Information Science & Technology (ASIS&T 2012), Baltimore, MD, USA, 26–31 October, pp. 83–95.

Ibekwe-SanJuan, F. (2012c). Information science in France: Emergence, evolution and perspectives. In A. Spink & J. Heinstrom (Eds.), *Library and information science trends in Europe* (pp. 273–295). London, UK: Emerald Library and Information science book series.

Ibekwe-SanJuan, F. (2012d). *La science de information. Origines, théories et paradigmes*. Paris: Hermès-Lavoisir.

Jeanneret, Y. (2007). Prendre en considération l'aventure sémiologique. In Perriault et Laulan (sous dir.), *Racines oubliées des sciences de la communication, Hermès*, *48*, 109–114.

Jeanneret, Y., & Ollivier, B. (2004). Introduction. Une discipline et l'université française. *Hermès*, *38*, 13–24.

Laulan, A.-M. (2007). Autour de Robert Escarpit: l'effervescence bordeaulaise (1960–1972). *Hermès*, *48*, 95–100.

Le Coadic, Y. (1994). *La Science de l'Information*. Paris: Presses Universitaires de France. (3ème ed.). 2004. Collection "*Que sais-je*?".

Maack, M. N. (2004). The lady and the antelope: Suzanne Briet's contribution to the French documentation movement. *Library Trends*, 53. Retrieved from https://pages.gseis.ucla.edu/faculty/maack/BrietPrePress.htm

Metzger, J.-P. (2002). Les trois pôles de la science de l'information. In Couzinet et Régimbeau (éd.) *Proceedings of the Colloquium "Recherches récentes en sciences de l'information : convergences et dynamiques"*, 21–22 March 2002, Toulouse, Paris: ADBS edition (pp. 17–28).

Meyriat, J. (1981a). Document, documentation, documentologie. *Schéma et schématisation*, 2e trimestre, *14*, 51–63.

Meyriat, J. (1981b). Information science soeur de la bibliologie. *Revue de bibliologie: Schéma et schématisation*, *15*, 9–19.

Meyriat, J. (1986). La science de l'information. *Revue des sciences morales et politiques*, *2*, 255–269.

Meyriat, J. (1993). Un siècle de documentation: la chose et le mot. *Documentaliste – Sciences de l'information*, juillet-octobre, *30*(4/5), 192–198.

Ministère de la Fonction Publique. Les Métiers de la Fonction Publique. Bibliothécaire: l'information à livre ouvert. Retrieved from https://www.fonction-publique.gouv.fr/archives/home20051107/fp/metiers/bibliothecaire.htm. Accessed on 14 December, 2017.

Palermiti, R., & Polity, Y. (2002). Dynamiques de l'institutionalisation sociale et cognitive des sciences de l'information. In R. Boure (Eds.), *Les origines des sciences de l'information et de la communication : regards croisés* (pp. 95–123). Villeneuve-d'Ascq: Presses universitaires du septentrion.

Perriault, J. (2007). Le rôle de l'informatique dans la pensée en information et en communication. *Hermès*, *48*, 127–129.

Salaün, J. M. (1991). La fin de l'IST ? Histoire des politiques publiques en information scientifique et technique. Rapport de recherche CERSI-ENSSIB, 40 pages. Retrieved from http://www.enssib.fr/bibliotheque/documents/travaux/salaun.pdf

Salaün, J. M. (1993). Les sciences de l'information en question. Retrieved from http://www.archivesic.ccsd.cnrs.fr/docs/00/06/20/53/RTF/sic_00000048.rtf

Tétu, J.-F. (2002). Sur les origines littéraires des sciences de l'information et de la communication. In R. Boure (éd.), Les origines des sciences de l'information et de la communication : regards croisés. Villeneuve-d'Ascq: Presses universitaires du septentrion, 71–93. Retrieved from http://halshs.archives-ouvertes.fr/docs/00/39/74/07/HTML/

Vitalis, A. (2007). Actualités de Jacques Ellul: La communication dans le contexte d'une société technicienne. *revue Hermes*, *48*, 163–170.

Warner, J. (2017). Feeding two wolves: The human and the computational in document analysis. In 11th Colloquium of the French Chapter of ISKO on epistemological and theoretical foundations of information science. Honouring francophone pioneers, Paris 11–12 July 2017, 8 pages.

Whitley, R. (1974). Cognitive and social institutionalization of scientific specialities and research areas. In R. Withley (Eds.), Social process of scientific development (pp. 69–95). London: Routledge & Kegan.

Yugoslavia

Atherton, P. (1965). Letter to the editor. *American Documentation*, April, 126.

Bauer, B. (1980). In memoriam of Bozo Težak. *Informatologia Yugoslavica*, *12*(3–4), 13–15.

Bauer, I. (2010). International permanent exhibition of publications in the light of Božo Tezak's contribution. In *Professor Božo Težak, Torchbearer of Science, Djurdjica Težak and collaborators* (Eds.), Croatian University Press and Informatics Association Božo Težak, pp. 217–224.

Čveljo, K. (1972). Educating librarians, documentalists and information scientists at the post graduate level in Yugoslavia. *Informatologia Yugoslavica*, *5*(1–4), 35–86.

Hayes, R. (2010). Professor Božo Težak, memories of a dear friend. In *Professor Božo Težak, Torchbearer of Science, Djurdjica Težak and collaborators* (Eds.), Croatian University Press and Informatics Association Božo Težak, pp. 346–349.

Merton, R. K. (Ed.). (1973 [1942]). The normative structure of science. In *The sociology of science: Theoretical and empirical investigations*. Chicago, IL: University of Chicago Press.

Mihel, I. (1989). From the referral center in Zagreb to the Institute of Information Sciences (IS). *Informatologia Yugoslavica*, *21*(1–2), 7–13.

Muljević, V. (1980). Božo Težak: Founder of the referral center at the University in Zagreb. *Informatologia Yugoslavica*, *12*(3–4), 35–41.

Muljević, V. (1989). Foundation and beginning of the referral centre of the University of Zagreb. *Informatologia Yugoslavica*, *21*(1–2), 1–5.

Muljević, V. (2010). A journey into European winter 1964/1965. In *Professor Božo Težak, Torchbearer of Science, Djurdjica Težak and collaborators* (Eds.), Croatian University Press and Informatics Association Božo Težak, pp. 248–252.

Plenković, M. (2010). Editorial world of the classic of information sciences—Božo Težak. In *Professor Božo Težak, Torchbearer of Science, Djurdjica Težak and collaborators* (Eds.), Croatian University Press and Informatics Association Božo Težak, pp. 313–320.

Prelog, N. (2010). Dubrovnik conferences – A net before the web. In Professor *Božo Težak, Torchbearer of Science, Djurdjica Težak and collaborators* (Eds.), Croatian University Press and Informatics Association Božo Težak, 341–345.

Sercar, T.-M. (2000). Toward an assessment of Bozo Težak's contribution to the development of information science. Invited Conference Paper. Retrieved from http://home.izum.si/cobiss/cobiss_obvestila/2000_2/Html/clanek_01.html

Sercar, T.-M. (2010). Težak's glocalism. In Djurdjica Težak and collaborators (Eds.), Croatian University Press and Informatics Association Božo Težak, pp. 228–294.

Skvorc, J. (2010). From the Institute for Scientific Information of Croatia to the interfaculty and the multifunctional experimental building and organization (MEZO): Unaccomplished projects that prepared us for the future. In Djurdjica Težak and collaborators (Eds.), Croatian University Press and Informatics Association Božo Težak, pp. 233–234.

Stearns, J. F. (2009). The National Referral Center. A new service in the Library of Congress. *Libri*, 1965, *15*(4), 353–359. Retrieved from https://www.degruyter.com/view/j/libr.1965.15.issue-1-4/libr.1965.15.1-4.353/libr.1965.15.1-4.353.xml

Stipanov, J. (2010). Bozo Težak: Visionary of the new national and university library building. In Djurdjica Težak and collaborators (Eds.), Croatian University Press and Informatics Association Božo Težak, pp. 229–233.

Težak, B. (1967). International permanent exhibition of publications – ISIP, Sveuc. Vjesnik (University Zagreb Inform.) 1959, 5, pp. 15–29 (Internacionalna stalna Izložba publikacija – ISIP). *Proceedings of the American Documentation Institute*, 4, 224–327.

Težak, B. (1969a). Introduction. *Informatologia Yugoslavica*, *1*(1–4), 1–2.

Težak, B. (1969b). Information, documentation and communication /INDOC / system. Emission – transmission – accumulation – selection and absorption (e-t-ac-s-a complex) as a conceptual basis of INDOC system. *Informatologia Yugoslavica*, *1*(1–4), 1–12.

Težak, B. (1969c). The role of small countries in international information systems with special respect to Yugoslavia. *Informatologia Yugoslavica, 1*(1–4), 31–41.

Težak, B. (1969d). Referral Center of the University of Zagreb. *Informatologia Yugoslavica, 1*(1–4), 65–77.

Težak, B. (1969e). Information, sciences and policies: Components, relationships and policies. *Informatologia Yugoslavica, 1*(1–4), 13–30.

Težak, B. (1970a). International permanent exhibition of publications – ISIP (Referral Center of the University of Zagreb) as a node point for an Information sharing network. *Informatologia Yugoslavica, 2*(1–4), 27–40.

Težak, B. (1970b). Curriculum and lecture program of courses at the center for the study of librarianship, documentation and information sciences. *Informatologia Yugoslavica, 2*(1–4), 95–185.

Težak, B. (1971). Informatics – The name for a confusing or clarifying concept. *Informatologia Yugoslavica*, 1971, *3*(1–4), 1–13.

Težak, Dj. (2010a). Quest for information systems and technology. In *Professor Božo Težak, Torchbearer of Science, Djurdjica Težak and collaborators* (Eds.), Croatian University Press and Informatics Association Bozo Težak, pp. 247–252.

Težak, Dj. (2010b). Fullbright grant in 1967 – Information era. *Ibid.*, pp. 252–254.

Težak, Dj. (2010c). Again America in 1971. *Ibid.*, pp. 261–262.

Težak, Dj. (2010d). The third information revolution: Internet. *Ibid.*, pp. 295–297.

Težak, Dj. (2010e). Unbroken trend of initiatives. *Ibid.*, pp. 301.

Težak, Dj. (2010f). International Referral Centre for Information Handling Equipment. *Ibid.*, 304–309.

Težak, Dj. (2010g). Library, institutes. Foundation of the Central Chemical Library, 1946. *Ibid.*, 84–87.

Težak, Dj. (2010h). Essential categories as assigned by Božo Težak. *Ibid.*, 3–10.

Tudjman, M. (2010). Božo Težak and the development of information science. *Ibid.*, 263–273.

Denmark, Sweden & Norway

Aabø, S., & Audunson, R. (2012). Use of library space and the library as place. *Library & Information Science Research, 34*(2), 138–149.

Aarek, H., Järvelin, K., Kajberg, L., Klasson, M., & Vakkari, P. (1992). Library and information science research in the Nordic countries 1965–89. In P. Vakkari & B. Cronin (Eds.), *Proceedings of the International Conference "Conceptions of library and Information Science" (COLIS). Historical, empirical and theoretical perspectives.* University of Tampere, Finland, 26–28 August 1991. London: Taylor Graham (pp. 28–49).

Almind, T. C., & Ingwersen, P. (1997). Informetric analyses on the world wide web: Methodological approaches to 'webometrics'. *Journal of Documentation, 53*(4), 404–426.

Audunson, R. (2005). Library and information science education: Is there a Nordic perspective. Paper presented at the World Library and Information Congress:

71st IFLA General Conference and Council, 14–18 August 2005, Oslo, Norway. Retrieved from www.ifla.org/IV/ifla71/papers/061e-Audunson.pdf

Audunson, R., & Aabø, S. (2013). From collections to connections: Building a revised platform for library and information science. *Information Research*, *18*(3). Retrieved from http://InformationR.net/ir/18-3/colis/paperC29.html

Audunson, R., & Gjestrum, L. (2012). Norway: Training of librarians in Oslo. *Scandinavian Library Quaterly*, *45*(3). Retrieved from http://slq.nu/?article=volume-45-no-3-2012-7

Audunson, R., & Gjestrum, L. (2015). Bibliotek og informasjonsfaglig utdanning: Fra etatsskole til iSchool. *Samle, formidle, dele: 75 år med bibliotekarutdanning*. Audunson, R. (red). Oslo: ABM-Media, 11–46.

Bubenko, J., & Dahlstrand, I. (2003). An interview with Björje Langefors. In J. Bubenko, J. Impagliazzo, & A. Solvberg (Eds.), *The history of Nordic computing* (pp. 7–15). Trondheim, Norway.

Danton, J. P. (1957). *United States influence on Norwegian librarianship, 1890–1940*. Berkeley, CA: University of California Press. Retrieved from https://babel.hathitrust.org/cgi/pt?id=mdp.39015022228616;view=1up;seq=13

Hannesdóttir, S. K. (2000). NORDINFO: Research and academic library cooperation across national borders. *Journal of Southern Academic and Special Librarianship*, Retrieved from http://southernlibrarianship.icaap.org/content/v02n01/hannesdottir_s01.html. Accessed on July 1, 2017

Harbo, O. (2000). Towards a Danish LIS university. *Journal of Documentation*, *56*(1), 5–11.

Hauk, K., & Stock, W. G. (2012). Pioneers of information science in Europe. The œuvre of Norbert Heinrichs. In T. Carbo & T. B. Hahn (Eds.), *International perspectives on the history of information science* (pp. 151–162). Medford, NJ: Information Today.

Ingwersen, P. (1994). The human approach to information science and management: The framework and prospects underlying the new Danish MSC program. *Journal of Information Science*, *20*(3), 197–208.

Ingwersen, P. (2004). Highlights of a career in information science. *Bulletin of the American Society for Information Science & Technology*, Aug/sept, 19–20.

Kajberg, L. (1988). The Royal School of Librarianship in Denmark. *Journal of Education for Library and Information Science*, *29*(2), 145–150.

Kajberg, L. (1991). Library and information science research in Denmark 1965–1989: A content analysis of R and D publications. *International Journal of Information and Library Research*, *3*(3), 187–220.

Katz, W. A. (2001a). *Introduction to reference work, Vol. 1: Basic information services*. (8th Ed.). New York, NY: McGraw-Hill.

Katz, W. A. (2001b). *Introduction to reference work, Vol. 2: Reference services and reference processes*. (8th Ed.). New York, NY: McGraw-Hill.

Limberg, L. (2017). Synthesizing or diversifying library and information science. Sketching past achievements, current happenings and future prospects, with an interest in including or excluding approaches. Keynote. Proceedings of the Ninth International Conference on Conceptions of Library and Information Science,

Uppsala, Sweden, June 27–29, 2016. *Information Research*, 22(1). Retrieved from http://www.informationr.net/ir/22-1/colis/colis1600.html

Lööf, S. (2000). Research, development of the discipline and a changing profession. Central issues for schools of library and information studies in Sweden. *Bibliothek Forschung und Praxis*, *24*(1), 36–43.

Lund, N. W. (2007). Building a discipline, creating a profession. An essay on the childhood of "Dokvit". In R. Skare, N. W. Lund, & A. Vårheim (Eds.), *Document (re)turn. Contributions from a research field in transition* (pp. 11–26). Frankfurt: Peter Lang Publishing Group. Retrieved from https://web.archive.org/web/20170814040544/http://munin.uit.no/bitstream/handle/10037/966/paper.pdf?sequence=1

Lund, N. W. (2009). Document theory. *Annual Review of Information Science and Technology (ARIST)*, *43*, 399–432.

Lund, N. W. (2016). How it all started: 1996, the first year of Dokvit. *Proceedings of the Document Academy*, *3*(1), 1–17.

Lundberg Rodin, M. (2012). Sweden: The librarian's profession is constantly changing. BHS turns 40. *Scandinavian Library Quaterly*, *45*(3), http://slq.nu/?article=volume-45-no-3-2012-5

Mäkinen, I. (2007). From marginal to excellence: The development of the research in information studies in Finland. In E. D. Garten, D. E. Williams, J. M. Nyce, & S. Talja (Eds.), *Advances in Library Administration and Organization* (Vol. 25, pp. 155–174). Amsterdam: Elsevier JAI Press.

Mäkinen, I., Järvelin, K., Savolainen, R., & Sormunen, E. (2016). From library and information science through information studies to information studies and interactive media: Emergence, expansion and integration of information studies at the University of Tampere illustrated in word clouds. *Information Research*, 21(1) March 2016. Retrieved from http://InformationR.net/ir/21-1/memo/memo4.html

Ørom, A. (2000). Information science, historical changes and social aspects: A Nordic outlook. *Journal of Documentation*, *56*(1), 12–26.

Persson, O., & Åström, F. (2005). Most cited universities and authors in library & information science 1990–2004. Retrieved from https://web.archive.org/web/20051120201815/http://www8.umu.se/inforsk/BibliometricNotes/BN2-2005/BN2-2005.htm

Pors, N. O. (2000). The education of information professionals in Denmark. *Bibliothek Forschung und Praxis*, *24*(1), 12–18.

Samuelson, K. (2006). Informatics and information systems – Including cybernetics and networks. In J. Bubenko, C. G. Jansson, A. Kollerbaur, T. Ohlin, & L. Yngström (Eds.), *ICT for people, 40 years of academic development in Stockholm* (pp. 196–200). Stockholm: University of Stockholm.

Tell, B. V. (1965). A Swedish course in scientific information transfer and retrieval. In *Proceedings of the 31st Meeting and Congress of the International Federation for Documentation, in cooperation with the American Documentation Institute*, Washington, DC, USA, October 7–16, 1965 (pp. 41–47).

Spain and Portugal

Abadal, E. (1992). *Origens i desenvolupament de la informacio i documentacio à l'estat espanyol: El problema de la denominacio*. Bellaterra: Publicacions de la Universitat Autonoma de Barcelona. (Doctoral thesis).

Caridad Sebastían, M., & Portela, P. (1989). The documentalist in Spain. *Education for Information*, *7*, 273–280.

Caridad Sebastían, M., Méndez Rodríguez, E. M., & Rodríguez Mateos, D. (2001). Information policies in Spain: Towards the new information society. *Libri*, *51*(1), 49–60.

Correia, A. M. R., & Wilson, T. D. (1992). The MSc in information management of the University of Sheffield taught in Portugal: An example of knowledge transfer in education. *Journal of Information Science*, *18*(1), 77–82.

Lopes Yepes, J. (2001). La investigacion Espanola en Teoria de la Documentacion (1990–2000). *Revista General de Informacion y Documentacion*, *11*(2), 259–280.

Lopes Yepes, J., & Sabelli, M. (2013). Library, documentation and information sciences in Ibero America. Towards a conceptualization and consensual definition. *Revue de l'Enssib*, *1*, 13 pages. Retrieved from http://bbf.enssib.fr/revue-enssib/consulter/revue-2013-01-001

Miralpeix, C., & Abadal, E. (2000). Education on library and information science in Spain. Development and current tendencies. *Bibliothek Forschung und Praxis*, *24*(1), 44–53.

Moreiro, J. A. (1989). *Agustín Millares Carlo: el hombre y el sabio*. Las Palmas: Cabildo insular.

Ortiz-Repiso Jimenez & Pacios. (2018). European trends in LIS education, as represented in WoS: 2000–2017. *Proceedings of the First International Symposium on the Future of Education in Information Science*, Pisa 10–11 September 2018 (pp. 153–164).

Portela, P. (1990). State of the art in education and training of library and information professionals in Spain. *Libri*, *40*(2), 126–134.

Ribeiro, F. (2007). An integrated perspective for professional education in libraries, Archives and Museums: A new paradigm, a new training model. *Journal of Education for Library and Information Science*, *48*(2), Spring 2007, 116–124.

Ribeiro, F. (2008). LIS education in Portugal between academia and practice. *Education for Information*, *26*, 33–42.

Rodríguez-Bravo, B., Simões, M., & Martínez-Ávila, D. (2018). The influence of documentation pioneer Paul Otlet on Spanish speaking and Portuguese speaking authors. In *Fondements epistemologiques et theoriques de la science de l'information-documentation: Actes du 11[e] colloque ISKO France 2017*, 11 et 12 juillet 2017. Widad Mustafa El Hadi (eds.). Paris, London: ISTE Editions (pp. 287–298).

Sabelli, M. (2010). Library and information sciences in the information disciplines environment: Towards integrative models of disciplines, professional community and information and communication public policies. In Unity in diversity, Proceedings of the Seventh International Conference on Conceptions of Library and Information Science (COLIS-7), *Information Research, 15*(4). Retrieved from http://informationr.net/ir/15-4/colis720.html

Sagredo Fernandez, F., & Garcia Moreno, A. (1998). History of information science in Spain: A selected bibliography. In T. B. Hahn & M. Buckland (Eds.), *Historical studies in information science* (pp. 263–268). Bedford, NJ: Information Today.

Others

Brezhneva, V., Gruzova, A., Krymskaya, A., & Soloveva, L. (2018). LIS education at the St. Petersburg State University of Culture: Trends and traditions. *Proceedings of the First International Symposium on the Future of Education in Information Science*, Pisa 10–11 September 2018 (pp. 16–25).

Cisek, S. (2002). Filozoficzne aspekty informacji naukowej, Kraków, Wydawnictwo Uniwersytetu Jagiellońskiego.

Dembowska, M. (1965). *Dokumentacja i informacja naukowa. Zarys problematyki i kierunki rozwoju*, Warszawa: SBP.

Hörzer & Schögl. (2018). Library education in Austria: With a particular focus on the Postgraduate University Program Library and Information Studies, MSc. *Proceedings of the First International Symposium on the Future of Education in Information Science (FEIS 2018)*, 10–11 September 2018, Pisa, Italy (pp. 94–102). Retrieved from http://einfose.ffos.hr/feis-2018/proceedings

Howard. (2018). Information Science: An emerging 'generalist' discipline? *Proceedings of the First International Symposium on the Future of Education in Information Science (FEIS 2018)*, 10–11 September 2018, Pisa, Italy (pp. 103–111). Retrieved from http://einfose.ffos.hr/feis-2018/proceedings

Index

Printed and bound by CPI Group (UK) Ltd, Croydon, CR0 4YY

13/09/2026

14961058-0001